NATURE'S REVENGE?

HURRICANES, FLOODS AND CLIMATE CHANGE

Institute of Ideas
Expanding the Boundaries of Public Debate

DEBATING MATTERS

NATURE'S REVENGE?

HURRICANES, FLOODS AND CLIMATE CHANGE

Institute of Ideas
Expanding the Boundaries of Public Debate

Tony Gilland
Mike Hulme
Peter Sammonds
Charles Secrett
Julian Morris

Hodder & Stoughton
A MEMBER OF THE HODDER HEADLINE GROUP

Orders: please contact Bookpoint Ltd, 130 Milton Park, Abingdon, Oxon
OX14 4SB. Telephone: (44) 01235 827720. Fax: (44) 01235 400454.
Lines are open from 9.00–6.00, Monday to Saturday, with a 24 hour message
answering service. Email address: orders@bookpoint.co.uk

British Library Cataloguing in Publication Data
A catalogue record for this title is available from
the British Library

ISBN 0 340 84840 5

First published 2002
Impression number   10 9 8 7 6 5 4 3 2 1
Year   2007 2006 2005 2004 2003 2002

Typeset by Transet Limited, Coventry, England
Printed in Great Britain for Hodder & Stoughton Educational, a division of
Hodder Headline Plc, 338 Euston Road, London NW1 3BH by Cox & Wyman,
Reading, Berks.

DEBATING MATTERS

CONTENTS

PREFACE
Claire Fox

NOTES ON THE CONTRIBUTORS

INTRODUCTION
Tony Gilland

Essay One   **CLIMATE CHANGE: A SOBER ASSESSMENT**
Mike Hulme

Essay Two   **EXTREME CLIMATES**
Peter Sammonds

Essay Three   **CLIMATE CHANGE: SCIENCE, POLITICS AND ECONOMICS FOR THE TWENTY-FIRST CENTURY**
Charles Secrett

Essay Four   **COPING WITH CHANGE: INSTITUTIONS FOR HUMAN HABITATION OF PLANET EARTH**
Julian Morris

AFTERWORD
Tony Gilland

PREFACE

Since the summer of 2000, the Institute of Ideas (IoI) has organized a wide range of live debates, conferences and salons on issues of the day. The success of these events indicates a thirst for intelligent debate that goes beyond the headline or the sound-bite. The IoI was delighted to be approached by Hodder & Stoughton, with a proposal for a set of books modelled on this kind of debate. The *Debating Matters* series is the result and reflects the Institute's commitment to opening up discussions on issues which are often talked about in the public realm, but rarely interrogated outside academia, government committee or specialist milieu. Each book comprises a set of essays, which address one of four themes: law, science, society and the arts and media.

Our aim is to avoid approaching questions in too black and white a way. Instead, in each book, essayists will give voice to the various sides of the debate on contentious contemporary issues, in a readable style. Sometimes approaches will overlap, but from different perspectives and some contributors may not take a 'for or against' stance, but simply present the evidence dispassionately.

Debating Matters dwells on key issues that have emerged as concerns over the last few years, but which represent more than short-lived fads. For example, anxieties about the problem of 'designer babies', discussed in one book in this series, have risen over the past decade. But further scientific developments in reproductive technology, accompanied by a widespread cultural distrust of the implications of these developments, means the debate about 'designer babies' is set to continue. Similarly, preoccupations with the weather may hit the news at times of flooding or extreme weather conditions, but the underlying concern about global warming and the idea that man's intervention into nature is causing the world harm, addressed in this book, is an enduring theme in contemporary culture.

At the heart of the series is the recognition that in today's culture, debate is too frequently sidelined. So-called political correctness has ruled out too many issues as inappropriate for debate. The oft noted 'dumbing down' of culture and education has taken its toll on intelligent and challenging public discussion. In the House of Commons, and in politics more generally, exchanges of views are downgraded in favour of consensus and arguments over matters of principle are a rarity. In our universities, current relativist orthodoxy celebrates all views as equal as though there are no arguments to win. Whatever the cause, many in academia bemoan the loss of the vibrant contestation and robust refutation of ideas in seminars, lecture halls and research papers. Trends in the media have led to more 'reality TV', than TV debates about real issues and newspapers favour the personal column rather than the extended polemical essay. All these trends and more have had a chilling effect on debate.

But for society in general, and for individuals within it, the need for a robust intellectual approach to major issues of our day is essential. The *Debating Matters* series is one contribution to encouraging contest about ideas, so vital if we are to understand the world and play a part in shaping its future. You may not agree with all the essays in the *Debating Matters* series and you may not find all your questions answered or all your intellectual curiosity sated, but we hope you will find the essays stimulating, thought provoking and a spur to carrying on the debate long after you have closed the book.

Claire Fox, Director, Institute of Ideas

NOTES ON THE CONTRIBUTORS

Tony Gilland is the Science and Society Director at the Institute of Ideas. He co-directed the Institute of Ideas' and New School University's Science, Knowledge and Humanity conference held in New York in 2001 and the Institute of Ideas' and Royal Institution's Interrogating the Precautionary Principle conference held in London in 2000. He is the editor of the Institute's Conversation in Print, 'What is it to be Human? – What science can and cannot tell us' (2001). Gilland holds a degree in philosophy, politics and economics from the University of Oxford.

Mike Hulme is Executive Director of the Tyndall Centre for Climate Change Research and is based in the School of Environmental Sciences at the University of East Anglia. His research specializes in the construction and application of climate change scenarios for impact, adaptation and integrated assessment. He has also worked on the evaluation of climate models, on the development of global and national observational climate datasets and on African climate and desertification. Hulme was a co-ordinating lead author for the chapter on 'Climate scenario development' for the Third Assessment Report of the United Nations Intergovernmental Panel on Climate Change (IPCC) as well as a contributing author for several other chapters. He has prepared climate scenarios for the UK Government, the European Commission, UNEP, WWF International and the IPCC. He writes a regular monthly climate column for *The Guardian*.

Julian Morris is Director of International Policy Network, a charity, based in London, that helps groups in developing countries in their struggle for democracy and freedom. He is also a research fellow at the Institute of Economic Affairs, a visiting professor at the University of Buckingham and a visiting fellow at the Liberty Institute in New Delhi. Educated in Edinburgh, London and Cambridge, Morris has degrees in economics and law. He is the author/editor of numerous books and monographs, including *Rethinking Risk and the Precautionary Principle* (2000).

Peter Sammonds is Professor of Geophysics at University College London where he directs the Bachelor and Masters of Science Geophysics degree programmes. He was a Royal Society Research Fellow for ten years and has been Visiting Professor at the University of Tokyo. He is author of some 40 scientific papers on a broad range of environmental topics, from the physics of natural hazards to ice sheets and is a frequent invited speaker at major scientific conferences, such as the 2001 American Geophysical Union meeting in San Francisco. He is a member of the Royal Society's Environmental Network and participates in the UNED Climate and Energy Group.

Charles Secrett has been Executive Director of Friends of the Earth since 1993. He is a member of the Government's Commission for Sustainable Development, and sits on the advisory boards for *The Ecologist* magazine, the Environmental Law Foundation and the Environment Programme of the University of North Carolina. His publications include *Adapting to Climate Change in Forest Based Land Use Systems: A Guide to Strategy* (1996); the annual 'Blueprint for a Green Economy' (FOE, 1995–2001 with Tim Jenkins); and *Making the Environment Work: Jobs and Environmental Investment* (1999).

INTRODUCTION
Tony Gilland

In recent years, the media and campaign groups have highlighted unusual weather events – from extensive floods in Mozambique and in parts of Britain, to droughts in India and hurricanes in America – as evidence of the growing dangers presented by climate change. Global warming is probably the most high profile environmental issue of our times. It is an issue which many governments are taking very seriously, although it is also a source of great controversy, especially with regard to debates about the future direction of economic development and humankind's relationship to the natural world.

Global warming first became a high profile public issue, and taken more seriously by governments and policy makers, in the latter part of the 1980s. This was a time when a whole range of environmental issues were on the political agenda. Issues such as nuclear power, 'acid rain', the ozone layer and global warming began to receive increasing amounts of media attention and environmental campaign groups such as Friends of the Earth and Greenpeace enjoyed substantial increases in their membership numbers. In 1988 Margaret Thatcher, who had previously been dismissive of environmental concerns, famously told the United Kingdom's premier scientific body, the Royal Society, that she now thought it possible that the impact of human activity on the environment constituted an unwitting experiment with the system of the planet itself.

One issue that exerted particular influence on the early discussion of climate change was that of the ozone layer. In 1985 an article in the science journal *Nature* announced that a hole in the ozone layer had been discovered over the Antarctic. Research later confirmed that chlorofluorocarbons (CFCs), widely used in such things as refrigerators and spray cans, could break down the ozone layer. This thinning effect was of concern because the ozone layer filters the sun's powerful ultraviolet rays and also because it was seen as a significant insight into the impact human activity can have on the environment. By 1987 a protocol was signed in Montreal, which put in place an international agreement to reduce the consumption of CFCs.

When attention then switched to global warming, and the 'greenhouse effect' of carbon dioxide and other gases (a warming effect caused by additional particles of these gases trapping heat radiated from the Earth – see later), the idea of establishing an international agreement to reduce human emissions of these gases was soon on the political agenda. However, the implementation of an international agreement to reduce human emissions of carbon dioxide has proved far more contentious and difficult, in part because the social and economic consequences of doing so are far more significant that those for CFCs.

At the high-profile 1992 United Nations Earth Summit at Rio de Janeiro, which brought together world leaders to discuss environmental issues, the United Nations Framework Convention on Climate Change was established. This provides an overall policy framework for addressing climate change. Its stated ultimate objective is 'stabilization of greenhouse gas concentrations in the atmosphere at a level that would prevent dangerous anthropogenic [man-made] human-induced interference with the climate system'. The convention entered into force in March 1994 following ratification by 50 of its signatory parties.

◈ ◉ ◉
◉ ◉
◉ ◉ ◈ **THE KYOTO PROTOCOL**

Since then the international community has been attempting to reach agreement on more detailed and specific commitments by individual nations to meet various targets to reduce emissions of greenhouse gases. The basis for these agreements is the Kyoto protocol, agreed in Japan in December 1997, but yet to be ratified and come into force. Under the terms of the Protocol, the developed nations of the world commit themselves to reducing their collective emissions of the six key greenhouse gases by at least five per cent compared to 1990 levels, during the five-year period 2008–2012. Regional targets range from an average eight per cent cut for most of Europe and a seven per cent cut for the USA up to a maximum ten per cent increase for Iceland and an eight per cent increase for Australia. Developing countries are expected to produce emissions reports and to cooperate with the agreement, but are not required to commit to specific emission reduction targets.

Attempts to finalize the operational details of the protocol, so that countries would finally ratify their agreement to be bound by it, became particularly rancorous towards the end of 2000 and through 2001 – as Charles Secrett describes in his essay in this collection. Some of the major issues that have caused dispute are: whether, and to what extent, countries should be allowed to trade rights to emit carbon dioxide (so that one country can purchase the right to emit more carbon dioxide than its target level from another country which has reduced its emissions beyond their target level); what penalties should be imposed on those nations that do not meet their target level of emissions; and the extent to which land management to increase 'carbon sinks' (see later) should count towards a nation's emission reduction targets.

Given the different geographical and economic positions of the countries involved, and disagreement over to what extent global warming is a serious problem, it is perhaps not surprising that the discussions have been protracted. A high profile and major rift between the Government of the USA and the European Union occurred at the Hague in November 2000 when the USA, alone among the developed nations, refused to agree to the latest version of the protocol and the talks collapsed. The US Government argues that meeting the terms of the agreement will be worse for the US economy than the consequences of global warming – an argument that has been met with much condemnation from many European governments. Talks restarted in Bonn in July 2001, when all the developed nations apart from the USA came to agreement on the terms of the protocol to be ratified. However, some details are still to be finalized and only a few countries have ratified the protocol to date and so it is not currently in force. Nevertheless, many governments are already acting upon the protocol to reduce their country's emissions.

UN INTERGOVERNMENTAL PANEL ON CLIMATE CHANGE (IPCC)

Another important aspect to the international response to climate change is the work of the Intergovernmental Panel on Climate Change (IPCC). The IPCC was established in 1988 by the United Nations Environment Programme (UNEP) and the World Meteorological Organization (WMO). The panel does not conduct research but rather it uses published scientific and technical research to assess the current state of scientific understanding of climate change, the extent to which the climate is likely to change and the implications of this. The IPCC currently has three working groups: one to assess the scientific aspects of the climate system and climate change; one to assess the negative and positive consequences of climate change and the options for adapting to it; and one to assess options for limiting greenhouse gas emissions and mitigating climate change.

The IPCC produces periodic assessment reports on the state of knowledge of causes of climate change, its potential impacts and options for response strategies. The First Assessment Report was completed in 1990 and formed the basis for negotiations on the UN Framework Convention on Climate Change, the second was produced in 1995 and the latest in 2001. These reports are highly influential and feed into the international discussions between governments. One important aspect of the IPCC's reports is the use of scenarios. In 1992 the IPCC formulated six scenarios about population and economic growth, deforestation rates, energy supplies and what action could be taken by governments to reduce emissions. For its latest assessment report (2001) the IPCC has taken a more uncertain view of the future and has used four storylines to generate 40 potential scenarios.

CONTROVERSY

Debates about climate change and its implications are both complex and highly politically charged. The essays in this book explore both the science of climate change and how this interacts with the politics of the situation. Mike Hulme, assessing the current state of scientific understanding, argues that 'we now have the unforeseen, and probably unwanted, ability to alter the very nature of our global climate system.' Hulme counsels against 'catastrophic portraits of the future' but also argues that climate change 'forces us to think about the long-term consequences of our decisions – more so than we are used to thinking.'

Peter Sammonds examines our scientific understanding of extreme weather events and argues that 'the public consensus in the UK, that adverse effects are already evident with worse to come, rests on very shaky scientific foundations.' Sammonds is highly sceptical of attempts to cut carbon dioxide emissions drastically, concluding that 'nothing should be done which would curtail global economic growth. ... Otherwise we would be damaging our very ability to adapt to environmental changes.' By contrast, Charles Secrett argues that our understanding of climate change and human impacts upon it can lead to a more enlightened vision for future growth and is optimistic about the benefits renewable energy can bring to both the economy and the environment. Secrett concludes that 'the idea of a global environmental space to be shared economically and politically by all people on an equal basis is a revolutionary concept' which can lead to a new 'approach to living, working and playing on this planet that works for the individual and the community, and minimises risks for generations to come.'

In the final essay, Julian Morris argues that 'attempting to control the climate through restrictions on emissions of carbon dioxide and other greenhouse gases is not only undesirable, it is futile.' Morris emphasizes the adaptive capabilities of humans to cope with change and that climate change 'is very low down on the list' of the genuine problems faced by the people of the world. The remainder of this introduction will provide a brief outline of the scientific issues underpinning the public debate about climate change.

FACTORS AFFECTING CLIMATE CHANGE

The Earth's climate is dependent on its exchange of energy with the sun and outer space. The atmosphere, the oceans, the land surface, the ice sheets and the Earth's biosphere affect this exchange. Given the large number of interactions involved, understanding how the Earth's climate is determined is a complicated process.

GREENHOUSE GASES

Most attention has been paid to the impact of the so-called 'greenhouse' gases. Light and ultraviolet radiation from the sun penetrate the Earth's atmosphere and warm the Earth's surface. This energy radiates back out from the Earth's surface as infrared radiation, which is absorbed by substances such as carbon dioxide, methane, nitrous oxide and water vapour (greenhouse gases or GHGs). This trapped energy keeps the Earth's surface about 30°C warmer than it would otherwise be and is vital to life as we know it. However, the amount of GHGs in the atmosphere has increased over the last two centuries and, through their 'greenhouse' effect, are thought to have contributed to increased global temperatures. Carbon dioxide levels in the atmosphere have risen by about 32 per cent in the past 200 years, increasing from about 280 parts per million to 370 parts per million today and methane levels are thought to have doubled over the last 100 years.

Increased levels of GHGs are primarily the result of human activities connected with energy generation, transport and agriculture – such activities have increased significantly since the beginning of the industrial revolution and will continue to increase as economies develop. In debates about global warming and how we should respond to it, the level of carbon dioxide has received the most attention, not least because it accounts for about 60 per cent of the extra heat-trapping gases. However, while most agree that human produced greenhouse gases have had some warming impact on the climate, the question of the magnitude of this impact is more controversial.

OTHER FACTORS

A great many other factors affect climate change and interact with one another – which is why understanding the processes at work and predicting the future climate is difficult. Some of these other factors include:

- Water vapour held in the lower part of the atmosphere, which can increase when more water evaporates, also traps heat.

- Other particles in the atmosphere, such as sulfates from industrial activities or dust from volcanic eruptions, rather than trapping heat reflect sunlight before it hits the Earth's surface, and are thought to produce a cooling effect.

- Clouds can both warm the climate (through trapping heat) and cool it (through reflecting incoming solar radiation) depending on their height and other properties.

- Oceans have an important impact on climate as the atmosphere gains from or loses heat to the oceans' surface. Essentially oceans act as giant heat reservoirs and redistribute heat globally via their circulation patterns. For example, 'El Niño', is a periodic event in the Pacific Ocean in which sea temperatures rise sharply on the eastern side and have a strong influence on the weather patterns throughout the world.

- Solar radiation varies due to physical changes in the sun. For example, the number of sunspots – dark patches on the sun's surface (or more accurately its gaseous layer), which result from a localized fall in temperature – fluctuate over an 11-year cycle. The duration of sunspot cycles may have an impact on the Earth's temperature.

- Finally, carbon dioxide is both stored within vegetation and soils and exchanged between them and the atmosphere. Taken together, vegetation and soils hold about three times as much carbon as the atmosphere. Plants use carbon dioxide as a fertilizer through the process of photosynthesis. Though carbon dioxide is eventually returned to the atmosphere through respiration it can be retained in tree wood for decades and in soil organic matter for centuries. This has led to the idea of maximizing the uptake and storage of carbon dioxide (carbon sequestration) through the management of land use and the creation of what have been called 'carbon sinks'.

HISTORY OF THE EARTH'S SURFACE TEMPERATURE

Systematic and global measurements of the Earth's temperature have only been taken for about 150 years. These measurements show that average global surface temperatures have risen by about 0.6°C over the last 140 years. The two key periods for temperature increases during this time are from 1910 to 1945 and from 1975 to the present day.

To understand longer term variations in global temperatures, scientists use proxy indicators. For example, temperatures can be estimated by examining tree rings because trees grow wider in warmer weather. Drilled ice cores allow climate scientists to infer past temperatures by examining the proportion of heavy and light oxygens contained within the ice. A series of glacial/interglacial cycles have been identified over the past one million years (driven by changes in the Earth's orbit around the sun, the tilt of the Earth's axis and the precession of the equinoxes). We are currently living in the last interglacial period (referred to as the Holocene) that began about 10,000 years ago and caused huge increases in sea levels as the ice melted.

More recently, over the last 1,000 years, two periods are generally identified as being significant: the 'Medieval Warm Period' in the first part of the millennium and the 'Little Ice Age' which broadly stretched from 1400–1900.

CLIMATE MODELLING AND SCIENTIFIC UNCERTAINTIES

CLIMATE MODELLING

The interactions between the various factors that impact on the Earth's climate are highly complex. In order to test their theories of how these mechanisms work and interact, climate scientists rely heavily on complex models running on supercomputers referred to as atmosphere–ocean general circulation models (AOGCMS – often abbreviated to GCMs). These models use a large number of complex equations to represent the interactions between different climate processes. The models divide the Earth's atmosphere into grids and solve the equations for each half-hour of time being simulated or modelled – the models will typically be set to model periods of several hundred years. Obviously, the accuracy of the models will depend on the extent to which they mirror the actual processes at work (which in turn depends on the current scientific understanding of those processes) and the accuracy of the available data that are fed into the equations.

Based on their survey of what science can currently tell us about climate change and on various scenarios about economic and social development across the globe, the IPCC's latest report estimates that by the year 2100 average global surface temperatures are likely to be between 1.4 and 5.8°C warmer than they were in 1990. This very wide range of likely temperature increase is due to both the uncertainties involved with the future character of economic and social development and the uncertainties surrounding the processes of climate change. The last point will now be discussed briefly.

'FEEDBACKS'

Given the complexity of climate processes it is not surprising that there exist many feedback mechanisms. A change in one factor that impacts on the climate will interact with the many other factors outlined above. Some of these factors will amplify a particular change (positive feedback) while others will dampen it (negative feedback).

Some of these feedback mechanisms and their likely impact on future climate change are poorly understood and are the subject of scientific uncertainty and continued scientific research and investigation. For example, significant uncertainty surrounds the net impact on future climate change of reflective particles – such as sulfates and aerosols – that are generally thought to have a net cooling effect, but the magnitude of this is uncertain and the processes are poorly understood.

The impact of water vapour is another feedback mechanism of great importance to predictions of future climate change where the magnitude of its impact is uncertain. As the Earth heats up, more water evaporates and traps more heat thus amplifying the impact of warming (positive feedback). However, the magnitude of this impact is dependent on the ability of the lower part of the atmosphere to hold more water, which is related to the extent to which the lower atmosphere warms up as surface temperatures rise. The extent to which this lower part of the atmosphere is warming is currently unclear. If it turns out that there is little warming, then this would significantly reduce the impact of this positive feedback mechanism and therefore estimates of the likely increases in future surface temperatures. Conversely, larger amounts of warming in this part of the atmosphere will significantly amplify increases in the Earth's surface temperature.

Finally, the impact of climate change on cloud cover and how this then feedbacks into the climate system is regarded as one of the biggest uncertainties in the whole picture of climate change. According to the IPCC 'probably the greatest uncertainty in future projections of climate arises from clouds and their interaction with radiation ... The sign of the net cloud feedback is still a matter of uncertainty, and the various models exhibit a large spread.' In other words clouds may be a part of a positive (amplifying) or negative (dampening) feedback mechanism and the magnitude of any such impact is uncertain.

CONSEQUENCES OF CLIMATE CHANGE

The consequences of climate change in the future obviously depend in part on the extent to which temperatures actually rise and, as we have discussed, the precise magnitude of future climate change is uncertain. How humans and other biological organisms will respond to different degrees of warming is also uncertain. Much of the debate about the consequences of climate change has reasonably focused on the impact on humans and the scope for humans to respond to the effects of climate change – this is central to the debate in this book.

The impact of sea levels rising, for example, will depend very much on the extent to which the different countries likely to be affected can protect coastal populations with sea defences. The IPCC estimates that the global water level will rise by between 10 and 90 centimetres by 2100, compared with a rise of between 10 and 25 centimetres over the past century. Clearly, in low-lying coastal regions human populations will have to contend with a greater risk of flooding, but the extent to which flooding will actually occur will depend on the extent to which those populations are able to protect themselves with sea defences as well as the actual amount by which water levels actually increase. Concerns have also been raised about warmer climates being more conducive to insects that transmit disease such as the mosquito. Again the impact of this is also dependent on human responses to this situation, particularly in relation to the provision of medicine and the use of techniques to control insect populations.

The possibility of a greater incidence of extreme weather events – heat waves, droughts, storms and hurricanes – has recently been the subject of much media interest, particularly following the unusually high levels of rainfall that caused major flooding in Mozambique in 2000 and the United Kingdom in the same year. What will happen to the frequency and intensity of extreme weather events as the climate warms is, however, very uncertain. Peter Sammonds explores these uncertainties in some detail in his essay.

How society responds to climate change has extremely important implications for the future of social and economic development. For some, the potential threats posed by a warmer climate are regarded as necessitating significant action now, particularly to reduce the level of human-produced carbon dioxide emissions. For others, the economic consequences of such actions are regarded as being far more harmful to the future well-being of people than the possible impacts of climate change and a major hindrance to the ability of future generations to respond to the environmental changes that they encounter. We hope the essays in this book provide an insightful and stimulating contribution to this important debate.

Essay One

CLIMATE CHANGE: A SOBER ASSESSMENT
Mike Hulme

Climate change – or the more popular term 'global warming' – has undoubtedly become one of the environmental icons of our time. Over the last 10–15 years it has increasingly entered into the public consciousness, into international politics, into our school curricula, indeed into the language and fabric of our culture. The term itself – global warming – is associated with a diversity of images and connotations, a majority of them negative, and in some cases associated with the language of catastrophe. The term as used by some people is almost a surrogate for everything nasty that nature, especially the atmosphere, might throw at us. Thus global warming is variously associated with more hurricanes, floods, droughts, storms, coastal inundation, heat waves, famine, illness, habitat loss, mass migration – all the dimensions of adverse environmental management our fragile human psyche is most worried about.

Is such concern warranted? Are such connotations of global warming founded on a scientific understanding of the phenomenon? How culturally determined are these images of global warming? Should we be talking the language of catastrophe rather than the language of management? Is, indeed, climate change 'the greatest environmental threat facing the world'? This essay will address seven key questions associated with the science and politics of climate change:

- Is our climate changing?
- Are these changes unusual?
- Why is climate changing?
- How will our climate evolve in the future?
- What are the direct impacts of future climate change?
- What are the indirect impacts of future climate change?
- Is climate change 'the greatest environmental threat facing the world'?

The response to each question will consider what is well established using the conventions of the scientific enterprise and what is poorly known. In some cases the responses will examine the extent to which our knowledge is grounded in scientific analysis or where views and perspectives on climate change are culturally or politically constructed. The main thesis of this essay is that the unforeseen, and probably unwanted, ability of humans now to alter global climate is the culmination of our species' increasing impact on the planet. Ensuring a sustainable long-term relationship between climate and society will therefore require new mentalities about our responsibility to future generations and new institutions and modalities of global governance.

IS OUR CLIMATE CHANGING?

The single most important index of global climate is the average annual surface air temperature of the planet. We know that this index averaged about 14°C during the period from 1961 to 1990 and this is a convenient base value from which to measure past and future changes. This global average value of course hides wide geographical variations, from an annual average temperature of 35°C or more in parts of the Middle East, to −50°C or less in Antarctica.

This global index of climate has been constructed quite reliably as far back as 1860, drawing on millions of individual thermometer measurements from around the world, including over the oceans. The index shows that the world is now about 0.7°C warmer than it was 150 years ago, averaging about 14.3°C during the 1990s compared to about 13.6°C during the mid-nineteenth century. The five single warmest years in this record have all occurred since 1990, the ten warmest years since 1981. There is little doubt that this most representative of indices of global climate shows a long-term warming trend.

More recently, since 1979, it has been possible to derive another index of global climate using satellite technology. Using a series of polar-orbiting satellites the average annual temperature of the free atmosphere – the temperature of the air between three to eight kilometres above the Earth's surface – has been estimated. Although of course this is not where people live, it provides us with another picture of what has been happening to our planetary climate system. This index shows rather weaker warming since 1979 than the surface record and much has been made of this difference. Some have argued that the satellite record is more believable than the index based on thermometer readings. Others have argued, however, that little can be said about long-term trends from an index that is only 23 years old. Others again have said that these two indices are measuring different aspects of the world's climate and that therefore they are unlikely to agree precisely.

To complete the story about what is happening to our Earth's climate, it is also necessary to examine a range of other indicators. It has been observed that the majority of the world's mountain glaciers outside the polar regions are contracting. For example, it has been estimated that, with a continuation of current trends, the glaciers at the top of Mt Kilimanjaro, located near the equator in Tanzania, may well have disappeared entirely within 30 years. Both the extent and thickness of Arctic sea ice have contracted over recent decades, the average ice thickness during September now being about 40 per cent less than half a century ago. There are indicators, too, of a changing climate in some of our ecological indicators. The growing season in northern mid-latitudes is now seven to ten days longer than at the beginning of the twentieth century and, in the UK, trees are leafing earlier and some bird species are laying their eggs earlier. And our best estimate of the average level of the world's oceans suggests that sea level has risen by between 0.1 and 0.2 metres during the last 100 years, a faster rate of increase than has been experienced for at least one to two thousand years.

Taken together, these meteorological, geophysical and biological indicators provide convincing evidence that the world's climate is indeed warming and that there are already detectable consequences of this warming in many parts of the world and for different aspects of our natural environment. We are living in a world that not only is seeing an acceleration in the rate of technological, social and culture change, but also in a world whose planetary climate itself is changing. The conventional assumption that prevailed in most of our planning, design and management institutions during the twentieth century that climate can be treated as a static boundary condition for modern society is no longer valid.

ARE THESE CHANGES UNUSUAL?

Answering this question depends on one's perspective. In the context of a single human lifespan (70 years) then 'yes', these changes are unprecedented. It is also true that over the last few generations – 100 to 200 years – the world has not experienced changes in climate of this magnitude or extent. Putting these current changes in climate into a longer term context, a thousand years or more, requires us to estimate past global climate change using a variety of indirect measurement methods: tree rings, ice cores, ocean sediments, landscape evidence etc. It is almost certainly true that the world has not warmed at the current rate, neither has it been warmer than now, during the last millennium. Fairly robust estimates of, at least, northern hemisphere temperature during the last millennium have been made using tree rings and these estimates suggest the hemisphere is warmer now than it was, for example, in the early Middle Ages.

On longer timescales still, we know of course that the Earth's temperature recovered from a major glacial episode that climaxed about 18,000 years ago. The world probably warmed from about 9 to 10°C at this time to about 14 to 15°C over a period of about 9,000 years, most likely reaching its warmest about 9,000 years ago. Within this 9,000-year period of deglaciation, there were periods of more rapid climate change when, at least for large areas like the North Atlantic Basin, temperatures may have risen by 1 or 2°C over less than a century. It is much harder to know for sure exactly how rapidly the world as a whole warmed or cooled during this period known as the early Holocene. We *do* know that average sea-level rose progressively as the major continental ice sheets began to melt and that this rise in sea-level persisted for many thousands of years. The most rapid rises in sea-level may have reached about 0.5 to one metre per century during the early to mid-Holocene, but slowed thereafter. These changes in climate and sea-level were, of course, natural in origin and were related almost certainly to the changes in the character of the Earth's orbit around the sun, amplified by various feedback processes within the Earth's climate system.

So, on these geological timescales, we cannot say that the changes in climate we are now experiencing *are* unusual; or, at least, we might claim they are unusual, but not unprecedented. Yet there are two aspects of these changes that certainly *are* unprecedented and both of these give us some cause for concern. First, this warming of our climate is now occurring on a planet with more than six billion people seeking not only only survival, but also the fulfilment of growing aspirations for greater wealth and enjoyment. Fifteen thousand years ago, there were probably only a few hundred thousand humans living on the planet and their essential support systems were very different from those of today. The second important difference is that there is a strong likelihood that the warming being experienced today is itself being caused in large measure by those same six billion human beings and their immediate ancestors. This warming, in other words, is unlikely to be just a natural phenomenon of planet Earth.

WHY IS CLIMATE CHANGING?

The processes governing the world's climate system are complex, but may conveniently be divided into external and internal. The basic *external* driver for our climate is the solar energy received at the top of the atmosphere. Changes in this quantity, the way it is distributed across the atmosphere due to changes in the Earth's orbit or changes in the associated electromagnetic spectrum of the energy received will have effects on the surface climate of Earth. The glacial cycles referred to earlier were in large part driven by changes in the distribution of this energy, and shorter, smaller variations in climate, for example, over 11 or 22 years, have been related to oscillations in the total amount of received solar energy. Other external factors that can alter surface climate are changes in the composition of the Earth's atmosphere, whether due to natural events, for example, large volcanic eruptions can cool surface climate, or to human activities, changing the concentration of greenhouse gases in the global atmosphere.

How large an effect these external factors have on climate depends in large part on the *internal* mechanisms that couple together the atmosphere, oceans, ice caps and biosphere. The exchanges of moisture, energy and chemical species between these components of the system are multiple and complex and still not fully understood. Because different components of the system respond on different timescales, their interactions can lead to complex non-random and non-linear behaviour. For example, the oceans respond much more slowly to an external change than does the atmosphere, and the biosphere controls the flow of chemical species between components of the system at different rates.

To understand the interplay of these external and internal factors better, climate scientists have developed a series of sophisticated computer-based models of the Earth's climate system. These models, called global climate or general circulation models, derive from the first principles of physics as originally specified by Sir Isaac Newton. They are based, therefore, on the same principles as the models that weather forecasters use day in, day out, to bring us our forecasts. In recent years it has become possible to conduct various experiments with the Earth's climate system using such models to help us understand the relative importance of external versus internal mechanisms and the role of human interventions versus natural processes. For the first time, therefore, it has become possible to test out in a computer 'laboratory' the theoretical ideas of the nineteenth-century scientists such as Joseph Fourier, John Tyndall and Svente Arrhenius, that greenhouse gases in the atmosphere, such as carbon dioxide, act as a major control on global temperature.

These computer-based experiments – the closest we are ever likely to get to conducting 'true' experiments with the world's climate – have taught us a great deal over the last 20 years. They have shown that the most important greenhouse gas in the atmosphere is water vapour; that major volcanic eruptions can cool the surface temperature for three to four years by up to 0.5°C; that changes in solar energy associated with the sunspot cycle can induce short-term warmings and coolings of up to a few tenths of a degree Celsius; and that if the concentration of carbon dioxide in the atmosphere doubles, then the surface temperature of the planet is likely eventually to warm by between about 1.5°C and 4.5°C.

The critical question to answer using such models is, 'What has caused the warming of the last 150 years?' If it is basically a natural phenomenon then although we may be able to predict its future evolution we can actually do very little other than to adapt as best we can to such changes. If on the other hand the warming *is* related to human activities, then there is the prospect of our being able to alter the rate of future climate change, at least to some extent and if we so desire. This question therefore has been at the heart of most climate science conducted over the last 15 years.

The conclusion from these studies has strengthened as the years have gone by: the observed warming, especially that recorded over the last 30 years, can only be explained in our models when we consider the effect of rising concentrations of greenhouse gases that are influenced by human activities. Although some of the early twentieth-century warming would appear to be related to changing solar activity and although some of the nineteenth-century warming may be the result of natural internal processes in the climate system, the rate and magnitude of warming experienced since the 1970s can *only* be fully explained by the continuing rise in the concentration of carbon dioxide, methane and other human-related greenhouse gases. By our expanding use of carbon based fossil fuels, humans are now not only altering *local* environments and climates (we have been doing this for centuries) but we now have the unforeseen, and probably unwanted, ability to alter the very nature of our *global* climate system.

How confident are we in this knowledge? Science, of course, always progresses through questioning established wisdom and by a commitment to a continuing search for more complete and powerful theories. There are very few areas of science, if any, that can claim total and inviolable certainty, least of all in relation to environmental science. Unravelling and understanding the complex interactions of our Earth's climate and life support systems is no exception. Our knowledge remains conditional and we continually seek out refinements to our theories and, on occasion, a rewriting of them. There are new hypotheses to be investigated and tested, supported or rejected. Some of these may weaken or indeed strengthen further our conviction that we have detected a substantial human influence on the global climate system. For example, it has been proposed that the role of solar energy on global climate might be greater than previously thought due to an amplification process relating to cosmic ray flux and atmospheric clouds. This hypothesis needs thorough investigation. Equally, it has been argued that we have perhaps underestimated the size of the human-amplified greenhouse effect because it has been masked by other industrial pollutants such as sulphate aerosols. Again, clarifying such a proposition requires further scientific endeavour. Such uncertainties, or dimensions of incomplete knowledge, are not to be shunned or suppressed; such debate and scepticism is to be welcomed as a motivation for better and more thorough science.

Yet we cannot limit our role as scientists simply to a never-ending quest for greater and greater certainty. At some stage we have to judge whether what we currently know, or believe that we know, warrants an assessment of the risks that lie ahead of us. And we have to judge whether such risks are sufficiently large to alert society so that we may debate whether or not a thorough risk management strategy needs to be implemented. This stage in climate change science was reached about 12 years ago and since then a series of assessments, risk assessments one might say, have been prepared by the world's climate scientists under the auspices of the United Nations Intergovernmental Panel on Climate Change (IPCC). We now consider their salient conclusions.

◆ ● ●
● ● ●
● ● ◆ **HOW WILL OUR CLIMATE EVOLVE IN THE FUTURE?**

The starting point for predicting future climate is to predict changes in the key drivers of climate change. For our purposes, this means the future growth in the emissions of greenhouse gases. This growth rate is not predictable in the conventional sense of the word since the drivers of emissions are subject to influences from human behaviour, political change, technological innovation and so on. Instead what is done is to construct a series of 'scenarios'; plausible descriptions of the future expressed in terms, for example, of population distribution, economic growth, technological change, global governance. Each of these scenarios, or stories, has a range of possible greenhouse gas emissions profiles, some of them increasing rapidly, others increasing less rapidly and a few eventually stabilizing.

When these different emissions scenarios are inserted into models of the global climate system a range of future climates is simulated, part of the range clearly being due to the different scenarios assumed and the other part of the range being due to different climate models making different predictions even with the same emissions scenario. The range of future warming out to 2100 currently considered 'likely' by the IPCC is between 1.4 and 5.8°C (referenced to 1990 base year); using a further set of models this warming translates into a range of increases in global average sea-level from 0.09 to 0.88 metres. In fact, about half of these ranges are due to different assumptions about the future and about half the range is due to uncertainties in the climate and sea-level models.

The lower end of the predicted warming range represents about a doubling in the rate of warming experienced over the last century, whereas the upper end of the range represents about an eight-fold increase. Clearly the consequences for society of these different ends of the range are quite different. How then do we proceed to develop a sensible risk management strategy? Before considering this question, we need first briefly to examine the types of consequences these changes in climate and sea-level might have for society.

WHAT ARE THE DIRECT IMPACTS OF FUTURE CLIMATE CHANGE?

A warming of global climate will inevitably alter the characteristics of weather experienced in all regions and localities. The climate system is a fully integrated system and warming it by one or more degrees Celsius will alter, for example, the distribution and magnitude of precipitation over the Earth's surface, the frequency, severity and distribution of storms around the world and the nature of the thermal regimes; especially extreme heat and cold. The issue is not trying to identify which individual experienced weather event has been caused or influenced by global warming – from now on *all* our weather is influenced to a greater or lesser extent by the human alterations to the global atmosphere. The more important question to answer is can we predict and, if so, with what confidence, the future changes in regional and local weather characteristics that follow from this warming of climate? Estimating these changes in future weather characteristics is crucial for redesigning our resource management systems, our social institutions and our policy regulations to be responsive to the changes in climate that are ahead of us.

Many attempts have been made to estimate and quantify the future direct physical impacts of climate change on a range of environmental and social systems. These studies often focus on water resources, food productivity, forest distributions, coastal flooding, human health, energy demand, tourism and the like. These resource systems and human responses are all, to a greater or lesser degree, sensitive to climate and hence to climate change and it is entirely appropriate that we undertake work aimed at quantifying just what the scope of climate change impacts might be. And there are some potentially serious disruptions that may be induced by a warming climate; for example, greater risks of riverine and coastal flooding or changes in the distribution of certain disease vectors and pathogens.

It must also be recognized, however, that future changes in climate will occur in a world that will be changing in nearly every other dimension as well and probably changing more rapidly than in the past. Some of these changes are likely to exacerbate stresses or dislocations induced by climate change, but some of them may well ameliorate the envisaged climate impacts or provide opportunities for new forms of wealth creation or to manage new environments. Customers, communities, corporations and countries will not be passive observers of these changes in climate; they will be active agents, in some cases pre-empting changes in climate through precautionary adaptation and in most cases by, at least, reacting to changes in climate and altering behaviour patterns, investment plans and regulatory policies.

Some of the more dire predictions of the impacts of climate change, therefore, need to be taken with some caution, especially where no consideration has been given to the nature and rapidity of social change or the capacity of systems to adapt. In many senses, these studies lead to rather pedagogical statements that draw necessary attention to the potential dangers associated with climate change. But as predictions of future reality they are poorly founded if they do not recognize the reflexivity of the systems being analysed or the broader context of social change and development.

WHAT ARE THE INDIRECT EFFECTS OF FUTURE CLIMATE CHANGE?

Quite distinct from the possible direct physical impacts of climate change are what might be called the indirect effects of climate change. These are the changes to economic and social policy that follow from the adoption of some form of risk management strategy and the consequences of these policy changes for customers, communities and corporations. We are already seeing some of these indirect effects in the UK and other countries.

At a global level, climate risk management has taken on the form of an international convention, the UN Framework Convention on Climate Change. Most of the world's nations have signed this convention, which came into force in 1994. The ultimate objective of the convention is, 'to stabilise greenhouse gases concentrations in the atmosphere at a level that would prevent dangerous anthropogenic interference with the climate system.' Quite what level is implied by this objective is, of course, subject to political negotiation and there are a number of radically different approaches that could be taken to defining 'dangerous'. The text of the convention provides some further direction by stating that such stabilization should be achieved, 'within a time frame sufficient to allow ecosystems to adapt naturally to climate change, to ensure that food production is not threatened, and to enable economic development to proceed in a sustainable manner.' However, 'dangerous' climate change might also be defined according to the future viability of certain sovereign states, such as Pacific atoll nations, whose very existence might be threatened by sea-level rise.

If the ultimate objective of the climate risk management strategy is difficult to define, how any such objective will be achieved is even less clear. The Kyoto Protocol, drafted in 1997 but not yet ratified, contains the first attempt to lay out a set of procedures and mechanisms that could start the process of delivering on the objective, that is, on starting consciously to reduce greenhouse gas emissions below levels that would otherwise have occurred. Despite the tortuous path towards ratification and the obstacles in the path of achieving full global ownership, the terms of the protocol have already triggered a stream of actions and reactions at both national and corporate level that should be seen as indirect effects of climate change. In the UK, for example, these effects include the introduction in 2001 of a climate change levy, an energy tax on certain businesses, and the commencement early in 2002 of a carbon emissions trading regime. These policy measures are altering the way in which some businesses, especially those in the energy sector, assess their energy policy and long-term strategic objectives. They have also set in motion a reassessment by investment fund managers about which companies are most likely to offer a good return on investment in a future carbon-constrained world. In terms of business risk and management these are real and measurable effects of climate change.

This reflexivity of society to the real or perceived threat of climate change is, therefore, a clear demonstration of the co-evolution of climate and society. Future climate, at least in the long term, is not predetermined. We can actually shape the climate of future generations, if not our own, by the decisions we make in this next decade. In effect, the inadvertent biogeophysical experiment we are conducting with planetary climate, has resulted in a parallel semi-managed socio-technological experiment. This experiment is about whether we can consciously shape, globally, our long-term energy-technology-lifestyle paths in a way that minimizes the dangers posed by climate change and maximizes the opportunities to restructure our economies into more sustainable forms. Such a conscious global social experiment has not been attempted before and our institutions are poorly designed to manage it.

IS CLIMATE CHANGE 'THE GREATEST ENVIRONMENTAL THREAT FACING THE WORLD'?

Not in the conventional sense. Catastrophic portraits of the future suit some world views and some temperaments, but do not encourage sober debate. If climate change is framed in terms of risk assessment and risk management, then it is natural that there will be a wide range of perspectives on just how large a risk is posed by climate change and how exactly that risk should be managed. All societies contain a mixture of people and organizations, some of whom are risk takers, and some of whom are risk averse. In effect, the instrument of democratic politics is designed to arbitrate in society on a whole range of issues between these different groups of people.

The language of catastrophe often used in climate change debates also tends to detract from a wide range of other environmental dangers or development objectives, which may not have the benefit of apparently being 'predicted' by powerful natural science models. Climate scientists have, perhaps uniquely in the environmental sciences, been bold enough to make predictions about the future 100 years hence on the basis of computer models of the climate system. These models exclude human behaviour and society and therefore have the appearance of greater predictive power than is warranted. There are not many other examples in social, economic or political planning where predictions over the next 100 years are used as a basis for policy design. Indeed, the genre of 'futures books' has, of course, a ready market, laying out visions of the wonders technology will have delivered and how social behaviour will have changed, but even these exercises in crystal ball gazing usually only think out to 2050. The worlds portrayed are often unrecognizable from our own.

The significance of climate change is not that it somehow opens up for us a new ability to predict the long-term future with any greater certainty than humans have always had. We must always be alert to the distinction between 'predictions' and 'scenarios' of the future; predictions hold themselves up as being objectively 'true' and scenarios describe themselves explicitly as being 'not true'. Predictions draw attention to the power and cleverness of the people or models that made them; scenarios divert attention away from the authors and force the user to ask themselves whether they *want* the scenario to happen.

The true significance of climate change therefore is that it forces us to think about the long-term consequences of our decisions, more so than we are used to thinking. It also forces us to think about issues of global equity and justice, perhaps more powerfully, say, than the parallel moral issue of the indebtedness of developing countries, because among the richer nations of the world the self-interest in tackling climate change is more obvious than in the case of debt relief. And climate change and its management is, in the end, a moral issue. It requires the deployment of moral or ethical arguments not only to define what constitutes 'dangerous' climate change, but also to agree on the principles of burden sharing between nations and individuals about who should first take action to reduce greenhouse gas emissions and to judge the level of climate risk we should bequeath to future, unborn generations.

Climate change should not be seen as the greatest environmental threat to the planet; it should be seen as a powerful mobilizing agent that forces humanity to think co-operatively about the sustainability of its long-term future on this planet.

Essay Two

EXTREME CLIMATES
Peter Sammonds

Projections of the twenty-first century, for example, Stanley Kubrick's film *2001*, communicated a sense that, as long as there was no nuclear war, humanity would continue on its bold, humanistic mission looking to new horizons beyond the Earth. Such projections, while hardly without their problems, were confident, outward looking and viewed risk taking with some optimism. Looking back on 2001 we will see the USA, the world's most advanced nation, paralysed with fear over terrorism and other leading industrialized countries questioning their success and wealth because of fears of the impact that very wealth has brought on the global environment. Early in 2001 the United Nation's panel of the world's leading climate scientists, the IPCC (Intergovernmental Panel on Climate Change), predicted global temperatures would rise by between 1.4°C and 5.8°C by the end of the century as a consequence primarily of carbon dioxide (CO_2) emissions from the burning of fossil fuels in the developed world. The President of the Royal Society of London, Sir Robert May, led calls from 17 academies of sciences from around the world for 'precautionary limits' on CO_2 emissions. A report in the same issue of *Science* magazine as the call from the academies suggested that the ancient Mayan civilization of Central America had collapsed in the ninth century AD because of drought brought on by climate change. Sir Robert commented on the BBC Radio 4 *Today* programme that we have to take action now on CO_2 emissions if we do not wish to go the same way as the Mayans.

In both the USA and the UK there have been calls at the highest levels of government for the application of the 'precautionary principle' to climate change. The first comprehensive study on climate change in the USA, *Climate Change Impacts on the United States*, prepared for the US Government by the National Assessment Synthesis Team (NAST) and a report by the British House of Commons science and technology committee, *Scientific Advice on Climate Change*, have both made the call. The popular media have echoed these calls. There have been numerous television science documentaries, both serious and lightweight, on extreme climate and weather and the threats they pose, but with the same basic message – that something *has* to be done about climate change. Spectacular flooding in the south of England in the autumn and winter of 2000/01 brought to British TV screens images of biblical-scale floods, with young children struggling through floodwaters and pensioners trapped in their homes by rising water levels. We had barely started commiserating with the victims before environmental campaigners laid the blame at the door of anthropogenic-induced global warming: an argument quickly spouted by government ministers. Others went further in presenting the floods as a warning to us all. According to Madeleine Bunting, a *Guardian* newspaper columnist, the more floods we suffer, the better: at least then we would wake up to global warming and get serious about negotiating carbon dioxide emission controls. The global warming debate has adopted this strongly moralistic tone with hurricanes, floods, snow blizzards and deadly heat waves all being seen as 'nature's revenge'. Sir John Houghton FRS (Fellow of the Royal Society), one of the IPCC's senior scientists, has gone on the record in *The Times* newspaper arguing that climate change is a moral issue, and linking it to in his own Christian beliefs. The World Wildlife Fund International (WWF) has accused the USA of being 'morally in the Dark Ages' and of 'holding the whole world to ransom' because of 'the rejection by the USA of legislative controls on CO_2 emissions.'

The orthodoxy on climate change is that global warming poses a terrible threat to the planet and upholding that view is clearly now a measure by which the morality of nations will be judged.

The solutions proposed to mitigate the effects of global warming range from dramatic changes in lifestyle to de-industrialization and population control. But one of the most widespread calls has been for 'demand-side restraint' on consumers through taxation, to moderate global warming. For instance, this has been argued for alike by the UK's last Conservative government to justify introducing tax on heating fuel and by the European parliament for a special tax on aviation fuel. Whenever objections are made to employing demand-side restraint, the images of extreme climates, hurricane damage, floods and droughts are brought out to put us back into the correct moral framework. For instance Paul Hyett, the President of the Royal Institute of British Architects writing in the *Architects' Journal,* targets the pursuit of individual freedom and the inability to curtail the excessive demands of free consumers as spelling doom for the planet. In support he cites a recent prediction that much of eastern England will be under the sea in 200 years. Bjørn Lomborg's book *The Skeptical Environmentalist: Measuring the Real State of the World* (2001) which argues, on the basis of an economic cost-benefit analysis, for minimal intervention in the environment, and against demand-side restraint, receives some praise in *Science* magazine particularly for its impressive breadth of coverage and statistical detail. But a review in the magazine *Nature* lumps this book together morally with Holocaust denial. There is a serious problem that the starting point for discussion on the environment has tended to not be considered scientific knowledge but alarmist interpretations of future climate set in a highly moralistic framework.

The IPCC distinguishes between what it calls 'simple' and 'complex' extreme events. The UN advisors are confident that the Earth's surface warmed over the last century and will warm over this century. There will be more days of extreme heat and fewer days of extreme cold. We can also expect increased precipitation globally and heavier rainfall. These are examples of simple extreme events. But what about complex extreme events such as storms and hurricanes? Most pundits and politicians take for granted that a warmer climate will lead to more complex extreme events. This is, however, not backed up by the IPCC. On the issue of storm activity and drought, it says in its 2001 report *Summary for Policymakers* that observed variation shows 'no significant trends evident over the last century.' As for future complex events the situation is even more difficult. The aim of this essay is to take a critical look at the evidence behind the global warming projections and the adverse consequences. I will argue that the public consensus in the UK, that adverse effects are already evident with worse to come, rests on very shaky scientific foundations.

ANTHROPOGENIC EFFECTS ON CLIMATE CHANGE

When one admits that nothing is certain one must, I think, also add that some things are more nearly certain than others.

Bertrand Russell, *Am I an Atheist or Agnostic? A Plea for Tolerance in the Face of New Dogmas*, (1947)

Global warming is happening. The IPCC puts climate warming over the twentieth century between 0.4°C and 0.8°C. In its *Summary for Policymakers* the IPCC states that 'most of the observed warming over the last 50 years is likely to have been due to the increase in greenhouse gas concentrations.' Computer models of ocean and atmospheric circulation projected back in time support this. Atmospheric content of the principal greenhouse gas carbon dioxide has increased from pre-industrial levels of approximately 280 ppmv (parts per million volume) to the current level of around 360 ppmv. (CO_2 is the most important greenhouse gas because its residence time in the atmosphere is about 30 to 80 years.) The IPCC also cites as evidence observational studies that show the intensification of El Niño (the warming of eastern tropical Pacific Ocean), the thinning of polar ice caps and warming of the world's oceans. Few authoritative voices doubt that global warming should not be taken seriously. However the key question is whether there is going to be a benign warming of up to 2 to 3°C or more extreme warming.

By a benign warming I mean returning to climate conditions present earlier in our current interglacial period about 6,000 to 7,000 years ago in what geographers call the mid-Holocene thermal optimum. Global average temperatures were 2 to 3°C warmer than they are now. The world overall was a warmer place, with higher rainfall and more biomass. Global precipitation was nine per cent higher than today. The Sahara Desert, which was considerably more extensive during the previous Pleistocene ice age, effectively did not exist. The savannah, replacing the desert, was far more hospitable to human life. It was only with the general cooling after the Holocene thermal optimum that the Sahara Desert grew to its present size. There were some areas of increased aridity, around what is now Turkey and in the Rockies, but in general the warmer climate of the Holocene optimum created a more fertile and productive earth (N. Roberts, *The Holocene*, 1989). So some global warming is not necessarily a bad or dangerous thing.

Some scientists do remain cautious about attributing too much weight to the anthropogenic influence on the global temperature last century. The temperature rise was relatively small and it could be argued that most global warming in the last century was due to the northern hemisphere emerging from the Little Ice Age (Figure 1), with much of this warming accounted for by increasing output from the sun. There has been cooling too due to volcanic dust. In the past 30 years, the temperature rise is mainly due to anthropogenic effects. In other words, it is due to carbon dioxide in the atmosphere from the burning of fossil fuels. But the magnitude of this temperature rise is small as can be seen in Figure 1, even though there has been a 30 per cent increase in the CO_2 content of the atmosphere. So what kind of temperature increases should we anticipate in the future?

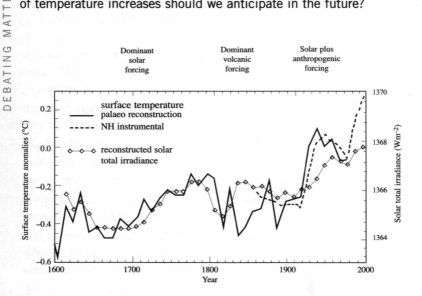

Figure 1
Comparison for 1600–1995 of the reconstructed solar total irradiance with the reconstructed northern hemisphere temperature record. The palaeo-temperature reconstruction uses tree ring growth data and instrumental data after 1860. Changing solar radiation and cooling due to volcanic dust can explain the long-term changes in temperature up to 1970 at which point anthropogenic forcing becomes important (From J. Lean, 'The Sun's radiation and its relevance for Earth', *Annual Review of Astronomy and Astrophysics*, 35 (33–67): 1997, Figure 10)

The IPCC in its 2001 report predicted temperatures would increase this century by between 1.4 and 5.8°C. In 1995 its prediction was for a rise between 1.5 and 3.5°C. Much has been made of the increase in the range of the predicted temperature rise, as if things were very much worse than we thought. The real difference is that in 2001 the IPCC considered scenarios where countries drastically cut emissions of sulphate aerosols (made to improve urban air quality), which form a cooling haze over parts of the world and considered different patterns of economic and social development. The actual projected *climate sensitivity* to CO_2 has changed little. The IPCC, although formally independent of the UN, works within a precautionary framework of the 1992 UN Framework Convention on Climate Change (UNFCCC), signed at the Rio Earth Summit. So the IPCC is doing nothing more than its job when it presents the worst that could happen, especially because of bad decisions. But many argue for a likely temperature rise towards the bottom of the IPCC projected range. For instance, William Dickinson, writing in *Eos, Transactions of the American Geophysical Union* (2000), argues that global warming due to the doubling or even tripling of CO_2 content in the atmosphere over pre-industrial levels would contribute a rise in temperature of between only 0.8 and 1.6°C by 2100: high projected temperature increases are a consequence of factoring in a large warming effect due to the build up of water vapour in the atmosphere. So the highest projected global temperature rises by the IPCC are a consequence of the double application of the precautionary principle both in regards of global socio-economic development and how the climate system works.

These figures for projected temperature rise are a global average. It is, of course, important to investigate the regional variations in this warming. The principal tools for climate modellers are the 'general circulation models' (GCMs) run on super-computers that couple the dynamics of the world's ocean to the atmosphere. One of the most advanced is the GCM of the Hadley Centre of the UK Met Office. The principal outputs of GCMs are maps of global changes in temperature and precipitation. Perhaps the most obvious feature of these projections is that most of the warming will occur at high latitudes. (There is also a strong contrast between winter and summer, with the winter months showing more warming than the summer months.) The landmass experiencing the most warming, according to the projections, will be Siberia in winter. Temperatures in cold Siberian anti-cyclones presently regularly drop below –40°C. So warming on a global scale will actually alleviate the worst weather conditions on Earth.

If the magnitude of likely temperature changes is uncertain, what are some of the impacts that might follow from higher temperatures? As I have commented, much is made of the possibility of a higher incidence of a variety of extreme events with very negative consequences associated with them. I now turn to examine these extreme events in detail.

SIMPLE EXTREME EVENTS

HEAT WAVES

Much has been made of the more frequent heat waves that will hit the UK with global warming. While a steady warming might be pleasant, wouldn't an increase in frequency of heat waves catch people unprepared? Are heat waves likely to be killers? Politicians cannot help but doom monger about the effects of global warming on health. Prime Minister Tony Blair, speaking at a WWF International conference in London on 6 March 2001, specifically raised the spectre of the increased risk of epidemics and the spread of disease with global warming with serious consequences for the nation's health.

Projected temperature changes for the UK produced by the UK Climate Impacts Programme (UKCIP) show that by the middle of the century even using a 'medium–high' emissions scenario, temperature changes for the UK, being a maritime climate, are not going to be large. Most years will be warmer than 1997 with its glorious weather. There will not be such a strong contrast between winter and summer temperatures as now. A warmer climate may only be a good thing for us. But what about the idea that the climate will become far more variable and prone to extreme heat waves? Temperature records used by the IPCC for the last 100 years show that there has actually been a *decline* in temperature variability within each year. There is also a tendency for lower variability in warmer years. The case that in general warm climates are stable and cold climates are unstable is strongly supported by evidence of past climates stored in deep ice cores from Greenland and the Antarctic (W.S.B. Paterson, *The Physics of Glaciers*, 1994).

However, establishing the impact of global warming on human health is clearly important. The UK Department of Health commissioned a report in 2001, *Health Effects of Climate Change*, about the United Kingdom. They found (using the UKCIP's 'medium–high emissions scenario' for 2050) that heat-related deaths occurring in the summer would increase by 2,800 per year. This represents an increase of 250 per cent! Even if weather-related deaths usually bring forward deaths by days to months rather than being the direct cause, surely this is ground for grave concern. However there must a strong suspicion that this number is a gross overestimate. It has been arrived at by assuming a threshold temperature band for heat-related deaths of 16–19°C and forward projecting the number of deaths by counting how many days there are going to be above this threshold. Even the report acknowledges the overestimate involved and states that 'health outcomes are sensitive to short-term meteorological extremes but are not likely to be significantly affected by long-term, incremental climate change.' As I argued earlier, temperature variability within each year could actually decline. Now contrast the position in winter. The UK has the highest 'winter seasonal excess mortality' in Europe, with an estimated 60,000 to 80,000 cold-related deaths. By 2050 this is projected to have declined by 20,000 per year due to global warming – a staggering drop. However, as the report points out, economic deprivation is an important factor in cold-related deaths as substantial numbers of elderly people live in 'fuel poverty'. Surely then, whether we are discussing extremes of heat in the future or cold in the present, the key issue is the ability of society to adapt to the prevailing climatic conditions.

There is unlikely to be any increase in epidemics in advanced countries. *Climate Change Impacts on the United States* states that, at present, much of the US population is protected against adverse health outcomes associated with weather and/or climate. There can be no more striking illustration of this than the incidence of dengue disease along the US–Mexico border. Dengue is a mosquito-borne viral disease. There is a striking contrast in the incidence of dengue in Texas, just 43 cases, versus the three Mexican states that border the USA where there were 50,333 cases in the period from 1980–96. This dramatically illustrates the importance of factors other than temperature, such as public health infrastructure and standard of living, on the transmission of diseases. There is no more reason to suppose that England would succumb to malaria.

A wet autumn in 2000 led UK Deputy Prime Minister John Prescott to warn that the floodwaters should be 'a wake-up call for everyone on global warming' as rivers burst their banks putting picturesque towns and villages under water. With global warming the Hadley Centre predict that rainfall will increase during the twenty-first century. Much of this global increase will occur over northern mid- to high latitudes. There may be increased precipitation in winter, but decreased precipitation in summer. The IPCC is also predicting large year-to-year variation in precipitation over most areas. There are claims that across northern Europe rainfall will increase, bringing with it the risk of flash floods. But according to the projections of the UKCIP's *Climate Change Scenarios for the United Kingdom*, a benign global warming would result in a modest increase in both annual and winter rainfall of just a few per cent and a drop in summer precipitation (particularly in the south-east of England). So what about the floods?

Periodic flooding surely is characteristic of an island with a maritime climate, where floods are a recurring theme. The total autumn precipitation for England and Wales from 1901 to 2000 is plotted in Figure 2 (prepared by Mark Saunders of the Benfield Greig Hazard Research Centre at University College London). Mean rainfall has remained remarkably constant, even though there has been warming throughout the century and particularly in the last 30 years. The wet autumn of 2000 is clearly an anomalous event: it falls into no pattern of increasing autumn precipitation or for that matter of wilder inter-annual oscillations in precipitation. There simply is not an increasing trend to these data.

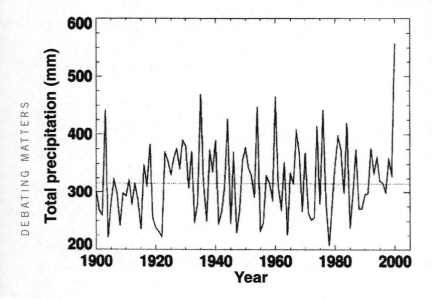

Figure 2
England and Wales autumn precipitation, 1901–2000 (M.A. Saunders, 'Climatic changes and their consequence for the reinsurance industry', invited presentation to the Benfield Greig European Seminar, Paris, 30 May 2001). Copyright retained by Dr Mark Saunders (University College London)

We must expect more flooding in the UK, however, for reasons unrelated to climate change. Increased urbanization means that more houses are built in river floodplains as more green fields are put under concrete and asphalt, resulting in faster water run-off. Another important factor are the attitudes of the UK Environment Agency and property developers (see N. Porcelli, 'How to stop the floods', www.spiked-online.com, 2000). The Environment Agency prefers soft 'green' flood defence measures and developers like 'natural' solutions. Contrast the leisure and amenity value of willow trees in greenfield settings, as opposed to grey concrete channels. But obviously these 'soft' solutions are less effective than the unpopular 'hard' solutions. The mechanisms that helped to save a large part of York from the worst affects of the recent floods were hard ugly structures. So how much flooding we are prepared to accept is far more closely bound to the amount of housing built in river floodplains and the type of flood defences we wish to build, rather than the degree of global warming.

DROUGHTS

> And then the dispossessed were drawn west – from Kansas, Oklahoma, Texas, New Mexico; from Nevada and Arkansas, families, tribes, dusted out, tractored out. Carloads, caravans, homeless and hungry; twenty thousand and fifty thousand and a hundred thousand and two hundred thousand.
>
> (John Steinbeck, *The Grapes of Wrath*, 1939)

There can be few things so powerful as the fear and devastation of drought. Floods subside, hurricanes blow themselves out, but drought destroys civilizations. It was this very powerful fear that Sir Robert May was playing to when he linked the possible collapse of the UN Kyoto Convention on Climate Change to the collapse of the Mayan civilization. According to the charity Christian Aid, global warming was responsible for the most severe drought in India for 100 years in the summer of 2000. 'One hundred million people in India,' the charity asserted, 'are paying the price of climate change.' In February 2001, CNN reported warnings of climate destruction and drought caused by global warming, resulting in the massive displacement of populations from the worst affected areas. These are indeed apocalyptic forecasts.

Even though the general outlook with global warming for the Earth might be positive, if we are not to go the same way as the Mayan civilization it is essential that our food supply be guaranteed. The USA is, of course, a major supplier of food to the world, accounting for more than 25 per cent of total global trade for some of the most important crops: wheat, corn and soyabeans. The forecast by NASA scientist James Hansen that droughts in North America would increase with global warming was instrumental in shaping the global warming debate and caused politicians to sit up and take note. However, the NAST report *Climate Change Impacts on the United States* specifically examined impacts on the agriculture sector. Increasing atmospheric concentrations of carbon dioxide has an important effect on crops as this generally results in higher photosynthesis rates (that is, has a fertilizing effect) and reduction in water loss. NAST examined the impact of global warming on US agriculture using the Hadley Centre model and Canadian Centre for Climate model. NAST projected changes in crop yields for 'dryland' (that is, CO_2 not irrigated) for 2030 and 2090. There are variations between crops, of course. The most striking features of these projections are the huge increases in crop yields that are expected to accompany climate change. So there does not seem much chance of us going the way of the Mayans!

COMPLEX EXTREME EVENTS

STORMS AND HURRICANES

Severe storms and hurricanes are the staple images of current TV weather news with usually some voice-over stating that worse is to come with global warming. WWF International in its backgrounder on climate change and hurricanes (*Global Warming and Atlantic Hurricanes*, 1999) states that if 'governments and corporations take more decisive action to reduce their carbon dioxide emissions they can minimise the risk of tomorrow's hurricanes becoming the *super storms* of the future.' They continue that it 'stands to reason that a warmer world will probably have more intense and more frequent tropical storms.' This ignores the fact that science is often counter-intuitive and has to be judged against empirical evidence. They argue that ocean surface temperatures above 27°C trigger hurricanes and every degree above this temperature produces an exponential increase in the propensity for storms. The insurance industry is as bad. The Loss Prevention Council in the UK states that losses, caused principally by windstorms, will be 'unimaginable' by the middle of the century. Gerhard Berz of Munich Re, the world's largest reinsurer, believes that 'there is reason to fear that climatic change will lead to natural catastrophes of hitherto unknown force and frequency', leading to worldwide losses 'totalling many hundreds of billions of dollars per year.' This talking up of the threat of global warming could lead to insurance companies dropping the risk posed by extreme weather events according to Julian Salt, an advisor to the insurance industry.

Where does this prediction of unimaginable losses due to storms and hurricanes in the warmer world come from? Figure 3 shows the reasoning behind the insurance industry statements. You assume a certain distribution of extreme weather events occurs with temperature (curve 1). You assume above a certain threshold temperature a number of these extreme weather events cause catastrophic damage – such as a large *landfalling* hurricane in the south-east United States. These assumptions, as discussed later, are contentious in themselves. But the truly staggering assumption is that with global warming you can just shift your entire distribution from position 1 to position 2 on the graph. *Eureka!* You have projected an astronomical increase in the number of extreme weather events causing catastrophic damage. But does this stand up to scrutiny?

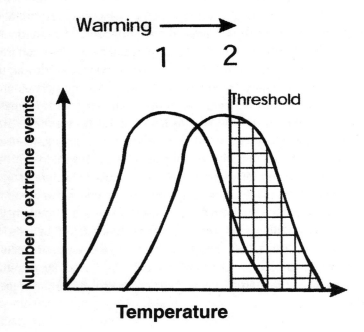

Figure 3
Schematic diagram of the number of extreme weather events such as windstorms plotted against temperature and supposedly how this changes with temperature

What will happen to the frequency and intensity of complex extreme events, such as hurricanes, with global warming is difficult to assess from climate models. Small-scale extreme weather such as tornadoes, thunderstorms and hailstorms are not even simulated in climate models according to the IPCC. There is evidence that the number of northern hemisphere *landfalling* hurricanes is increasing and that the proportion of tropical cyclones reaching intense hurricane strength is increasing. However, as Mark Saunders explains in his recent review of climate change in the *Philosophical Transactions of the Royal Society* (1999), these increases are a result of natural variability rather than due to global warming. The IPCC reports that there have been no systematic changes in the frequency of the small-scale weather extremes. Further, even if there were a propensity for an increase in the number of extreme weather events, this does not translate into actuality. The late Sir James Lighthill, former Provost of University College London, found in a major study that any possibility of increase in frequency or intensity of hurricanes was swamped by natural variability (*Bulletin of the American Meteorological Society*, 75, (2147–57): 1994). Science is frequently counter-intuitive. It is not inevitable as heat builds with the ocean–atmosphere system that this should result in more frequent violent releases of energy. With global warming, the higher latitudes will warm more than lower latitudes. As the temperature differences narrow between high and low latitudes the pressure drive will lessen. So as Patrick Michaels and Richard Balling explain in their book *The Satanic Gases* (2000) warmer winters in polar regions will produce a weaker jet stream with possibly fewer and less powerful cyclones, that is, a warmer climate could lead to fewer complex extreme events.

The general case for more violent atmospheric circulation with global warming is also not borne out by palaeo-climate data. The Pleistocene, or the last two million years, has been the period of ice ages. These were extensive worldwide glaciations lasting 100,000 to 150,000 years, punctuated by inter-glacial periods lasting

10,000 to 20,000 years. It is the present inter-glacial, the Holocene, which has seen the development of civilization. The temperature, CO_2 and dust records of past climate, trapped in deep ice cores taken from Greenland and Antarctica show that it is during ice ages there is high dust content in the ice core. This high dust concentration is indicative of high atmospheric circulation. During inter-glacial ages, when the climate is warmer, dust content is 40 times lower (W. S. B. Paterson, *The Physics of Glaciers*, 1994). So it seems it is the ice ages that have high atmospheric circulation, not warm periods.

EL NIÑO

Everybody is familiar with abnormal temperatures, rain or storms occurring every few years. The most important causes of this year-to-year natural variability in the world's climate are El Niño and La Niña events. In the Pacific region there are departures from the norm of atmospheric pressure, wind, temperature and cloudiness, which have a global impact. These departures coincide with the appearance of warm water off the coast of Peru in the case of El Niño events or cold water in the case of La Niña events. It is the El Niño events, that are blamed for the destructive weather around the globe. For instance, the 1997–98 El Niño is estimated to have caused $500 million of damage in California. The IPCC, in its 2001 report, says that more frequent, persistent and intense El Niños since the 1970s are symptomatic of global warming. WWF International in its backgrounder on hurricanes warns that the world could slip towards a semi-permanent El Niño state. But what are all the fears about?

The tropical trade winds in the Pacific usually blow from the eastern Pacific to the western Pacific moving the surface waters westward across the Pacific. Upwelling cold water off the coast of South America keeps the tropical water of the eastern Pacific a few degrees cooler than in the west. These trade winds slacken off seasonally. An El Niño event is when there is a persistent slackening or reversal of the trade winds, accompanied by the appearance of unusually warm water off the South American coast. La Niña is the opposite part of the cycle where unusually cold water appears off the South American coast. El Niño and La Niña are part of the seesaw in atmospheric pressure between the Pacific and Indian Ocean called the Southern Oscillation. About every four years abnormal patterns of temperature, rainfall and storminess occur around the globe due to the El Niño Southern Oscillation. Since 1980 there has been a trend towards more El Niños and fewer La Niñas. As Saunders explains in his review of climate change (*Philosophical Transactions of the Royal Society*, 1999), the magnitude of these climate anomalies currently exceeds the changes expected due to anthropogenic global warming for 2100.

Like all climatic events there are benefits as well as negative consequences. El Niño is associated with warmer winters in the Unites States bringing savings of $7.5 billion in winter fuel costs. El Niño weakens hurricanes in the Atlantic basin. El Niño brings much needed rain to southern California. In fact as Michaels and Balling point out, societies have adapted themselves to this natural variability, since the earliest humans had to endure El Niño induced droughts in East Africa (*The Satanic Gases*, 2000).

CONCLUSION

There is a strong public consensus that we may be facing disaster as a result of global warming. The idea that the Earth should operate in some kind of a balanced equilibrium is the foundation of this consensus – but global climate is always changing. Global temperatures are projected to continue to rise this century. However, assessments of the impacts of this temperature rise do not conform to a scenario of impending disaster. Projections for this century indicate a more productive agricultural sector. A warmer climate will see an overall improvement in health because of a steep decline in cold-related winter deaths. Adverse consequences for health of global warming would be limited by improving the public health infrastructure. There is little evidence for increasing complex extreme events such as super hurricanes. Natural disasters will continue to hit developing countries. But the answer for this has to be faster economic growth and technological advance.

Much has been said about the terrible floods in Mozambique in 2000. The immediate cause of the floods was a series of four cyclones, which hit the country between January and March. These caused the heaviest rainfall and worst floods for more than a century according to a Channel 4 report online. Most people in the area affected were peasant farmers. They lost their crops, their seeds and farm implements and were unable to plant again because the ground was still waterlogged. All the rice and sugar in that region was lost and extensive damage done to the irrigation systems. The underlying cause of the floods was attributed to La Niña. But the idea that during the course of the century Mozambique would not have the money for comprehensive flood defences or financial reserves to cope with an agricultural crisis says a lot about how the west views the likely development of Africa. It is at this juncture that the issues of global warming and poverty meet. It is argued by environmentalists that global warming is an additional stress on the system, already under pressure from population growth in a world of finite resources. However, if we reject the finite resources argument and, instead, argue that what is needed is more development to end poverty and not less, this also points the way to addressing the problems that could result from global warming. For instance, we must promote rapid industrialization and the application of advanced technologies for Mozambique and not check development through a misplaced sense of what is appropriate for the country. The IPCC global warming scenarios with the greatest projected temperature increases are also the ones with the greatest development and ability to develop technologies tackling certain pollutants, that is the ones accompanied by the greatest ease of adaptation.

It is interesting to return to Sir Robert May's point about not wanting to go the way of the ancient Mayan civilization. The report in *Science* on the collapse of the civilization in the ninth century AD argued that it was because of drought brought on by *increased activity of the sun* resulting in high levels of solar radiation reaching the Earth (Hodell *et al.*, *Science*, 292 (1367–70): 2001). This was a society that had not reached the level of social and economic development necessary to cope with environmental change. Neither was it part of a global economic system which could buttress it during bad times. It is the level of development and globalization that are the key factors. The lesson to be learned from the collapse of the Mayan civilization is not that our current levels of carbon dioxide emissions should be savagely cut; but rather nothing should be done which would curtail global economic growth through misplaced strategies to curb carbon dioxide emission. Otherwise, we would be damaging our very ability to adapt to environmental changes. This argument applies not only to countries such as the Mozambique, but also to the USA and other advanced countries, to which the world will continue to look this century because of their massive agricultural output and technological innovations. In the long term societies must be in control of their environments to create the optimum conditions for people to live in. But it is clear that significant economic and social development throughout this century will be needed before that can be achieved globally.

Essay Three

CLIMATE CHANGE: SCIENCE, POLITICS AND ECONOMICS FOR THE TWENTY-FIRST CENTURY

Charles Secrett

In July 2001, 178 governments agreed in Bonn to implement a global, legally binding concord (the so-called Kyoto Protocol) to reduce the pollution that is warming the Earth's atmosphere. This was an historic first: a truly unprecedented political response to the deep divisions between nations, which had marked 15 years of often bitter negotiations, as the world decided what it wanted to do about the emerging threats of climate change and sea-level rise.

So does this mean we can all rest easy because everything is done and dusted? Not at all. The arguments go on. There are some very dominant players who refuse to play ball. The USA, the Earth's worst climate polluter, its richest nation and strongest political economy, has rejected the protocol and will not sign up to emission reduction targets. Corporate giants like Exxon/Mobil and business lobby groups like the Global Climate Coalition, funded by oil, coal and motor multinationals, continue to oppose concerted action. At every climate summit, and there are many more to come, nations manoeuvre for special advantage, stretching negotiations to brinkmanship. Superficially, the disagreements are over the science of climate change and its reliability. Actually they are about power and control; nations argue over valuable resources, political decision making and market share.

The threat of serious climate change has imposed, for the first time, a planet-wide array of interlocking problems on society and a set of choices of unparalleled magnitude. Yet political leaders seem incapable of stepping back from familiar routines and considering how profound a challenge we face. What does the science really tell us? Do we respond to, or ignore, the warning signs that a major realignment of planetary ecosystems is occurring? Will a free market provide the best solutions or is coordinated global governance necessary? What does strategic security mean if the climate system is destabilized, weather patterns become unpredictable and extreme events such as hurricanes grow more intense and happen more frequently? Seeking answers, nations are having to reassess fundamental and cherished assumptions about the nature of progress, economic development and international relations.

At the heart of each of the significant environmental decisions we face, from energy and fuel use to forest management and resource consumption, lie power struggles and rivalries. Who gains? Who loses out? Who controls? Who profits? Who decides? The battle lines shift ceaselessly between communities, companies and countries; between those who demand reform and those who defend the status quo. During recent climate negotiations, as at World Trade Organization meetings and other summits, the poor, historically disadvantaged Southern countries have been standing up for their rights. They are determined to secure the political authority and economic resources – including a fair share of the climate's capacity to absorb pollution safely – required to meet their people's development needs. In Bonn, Europe agreed and enough compromises were made to ensure sufficient support from other reluctant industrial governments to do a deal. Despite the uncertainties, and fierce US opposition, the protocol was enacted and, unprecedentedly, America lost. How will President Bush and

Congress react? The established world order is beginning to break up and we cannot predict the consequences.

There are other reasons to stay alert. The Kyoto Protocol is just a small step toward implementing the United Nations Framework Convention on Climate Change, formally established at Earth Summit 1 in Rio de Janeiro in 1992. The targets set in Bonn for reducing greenhouse gas emissions such as carbon dioxide are nowhere near sufficient to stabilize pollution levels, and so prevent 'dangerous anthropogenic [sic, human] interference with the climate system', as the Convention sets out. The net effect will be to reduce carbon emissions by between one and three per cent on 1990 levels. The most robust scientific assessments, provided by the Intergovernmental Panel on Climate Change (IPCC), estimate that global cuts of at least 60 per cent by 2050 are needed to meet that goal.

For the first time in history, the combined activities of mankind are disturbing global ecosystems. The pollution impacts of energy, transport and land use choices made by households, businesses and government departments all over the planet have grown to the extent that the climate system can no longer absorb them without changing. Emissions of gases such as carbon and methane are altering the chemical and energy loads that the system is based on. Consequently, the relatively stable and predictable weather patterns we have come to rely on are taking new forms, as other knock-on effects like sea-level rise are simultaneously kick-started. These are the new ecological realities to which society must adjust.

The climate debates are not all about politics and economics. We are learning a lot about our wider relationships with nature and with ourselves. Throughout history, the weather has been something that happens to us, beyond the stretch of man. Now we are discovering that our connections with the sky are no longer passive and one-way. They are mutual and interactive. As always, the weather affects us; but now our behaviour affects weather patterns. There is another change, too; the relationship is no longer only local and immediate. Our lifestyles and the pollution we cause spark economic and social consequences which stretch across the global community and for future generations – from the 'here and now' to the 'there and then'. The political implications of personal choices made in millions of households have swollen accordingly. Perspectives like these open up an unfamiliar moral dimension to everyday decisions, ones usually taken without a second's thought. The most appropriate course of action to follow must now account for the needs of the global community of citizens – not just each patch separately – and for future generations, not just our own. And everyone, not just government or industry, has a part to play.

But how? And who should act first? How do we minimize the costs and maximize the benefits of adapting to climate change? Who should help whom? What time do we have? These are questions that the next summit rounds must address. The climate sceptics have had a good run for their money. But the protocol agreement effectively signalled the end of the doubters' agenda: as climate change is happening, will nations act together and regulate markets? Are governments generally prepared to pool sovereignty to make decisions? The answer from Bonn is yes, except for America.

THE SCIENCE AND THE LIKELY CONSEQUENCES OF CLIMATE CHANGE

The science of climate change can now tell us three things that we need to know. First, that the Earth is getting progressively warmer and is warming more quickly than at any time over the past 1,000 years. Second, that human beings are primarily responsible. Third, that complex systems, such as the climate, oceans, forests and economies, operate and interact in surprising, often unpredictable ways. These are facts.

The IPCC has recently updated its scientific, impact and policy assessments. They make sober reading. It is worth remembering that many of the 1,050 scientists on the panel work for or are funded by governments and corporations that have resisted early action to tackle carbon and other climate pollution. As the open and rigorous enquiries of the IPCC have progressed, previous uncertainties over climate change processes and likely outcomes are being pessimistically resolved. The IPCC now concludes that it is at least 90 per cent certain that the average global surface temperature of the Earth will warm between 1.4–5.8° Celsius by 2100. These increases are almost twice those predicted in their last report, some five years ago. Sea-levels are expected to rise, as the oceans in turn warm up and expand, within a 15–95 cm range. The panel has also hardened its conclusions about the relative contributions of natural climate variability and human behaviour to the observable warming. In 1996, they cautiously stated that the balance of evidence suggested a discernible human influence. Now they state that there is new and stronger evidence that most of the warming observed over the last 50 years is caused by people burning fossil fuels and forests.

This matters. Since the onset of the industrial revolution, concentrations of the main greenhouse gas, carbon dioxide (CO_2), have risen from about 285 parts per million (ppm) to 368 ppm today. The rate of increase in CO_2 is the highest for at least 20,000 years and present concentrations have not been exceeded for at least 420,000 years. If we continue to burn fossil fuels at current rates, atmospheric CO_2 will be twice pre-industrial levels by 2030 and three times by 2100. (A doubling of pre-industrial CO_2 levels, to between 450–550 ppm, is considered the upper limit of tolerable concentrations.) In temperature terms, the 1990s were the warmest decade in a millennium, and the anticipated increases are without precedent during the past 10,000 years.

The panel details the many possible threats to the natural world and society, and likely losses from the anticipated droughts, floods, sea-level rise and extreme weather events such as hurricanes, which accompany climate change. There are some potential, moderate climate gains, such as extended growing seasons, in a few places. The best thing to do is to read the original reports, published by Cambridge University Press.

Ecosystems as diverse as coral reefs and atolls, glaciers, mangrove swamps, northern and tropical forests, polar and alpine regions, wetlands and grasslands will undergo significant and irreversible damage unless emission cuts are made. In many cases, serious changes are already observable. Rare and endangered species will move closer to extinction and many are likely to disappear. Plants and animals have evolved under stable climate conditions for the past 10,000 years. Very few can adapt to such relatively rapid disruption.

For most humans, the outlook is equally dire. Take water, as just one of many problematic examples. In regions such as Europe, where supplies are adequate, more frequent, heavier rainfalls are expected, increasing the risk of disruptive river floods. Britain has not coped well with its recent heavy rain. But lower rainfall and water shortages are anticipated in arid areas like northern and sub-Saharan Africa, central Asia and Australia. Some 1.7 billion poor people live in these water-stressed regions. Droughts mean failed crops and food shortages. The IPCC predicts a general reduction in crop yields in most tropical and sub-tropical regions.

The panel warns of greatly increased risks from spreading insect-borne diseases, such as malaria and dengue fever and water-transmitted diseases, such as cholera. Increases in heat waves, exacerbated by humidity and air pollution, will lead to rises in heat-related deaths and illnesses in the north as in much of the South. People most at risk are the elderly, sick and young, living in towns and cities. With a mid-range sea-level rise of 40 cm and the anticipated increase in violent storms, the number of victims of coastal storm surges may to rise to 75–200 million, depending on preventive measures. Tens of millions of vulnerable people already live below sea-level in countries such as Bangladesh and Egypt. Climate change disruptions will cause large numbers of refugees in poor countries to flee afflicted areas.

If there is good news, it is this. The panel believes that the wildest speculations of tabloid headline writers are probably just that, if emissions are cut. The most dramatic, irreversible global system changes – such as a significant slowing of the Gulf Stream, the disintegration of the Greenland and Western Antarctic ice sheets or the release of massive carbon stocks locked up in permafrost regions and methane in coastal sediments – are unlikely to occur. But, if current warming causes additional permafrost and sediment releases, temperatures will rise more, leading to further releases, leading to increased temperatures, melting, ocean current changes and so on. That is the worst case scenario. The trouble is no one knows when or if temperatures will rise sufficiently to trigger feedbacks like these. On computers, climate–ocean circulation models do show how they are possible, but the projected outcomes are very dependent on which future climate scenario datasets are used. Nevertheless, the likelihood of one or all of these mega-scale events happening increases while the warming of the atmosphere and oceans goes on.

Just how dangerous can a few degrees hotter really be? Well, the present global temperature is only 3.5° Celsius warmer than during the last Ice Age, yet the planet is a completely different place to live in. As the IPCC warns, small changes in average climate conditions can kick off sudden shifts into significantly different climate states, and catalyze relatively large and unexpected changes in the occurrence and frequency of extreme climate events, on global and local scales. Unpredictability and flipstate switches are what uncertainty means when you are dealing with the behaviour of complex, chaotic systems, like the climate. Nasty surprises can come out of nowhere.

BATTLE LINES AT BONN

If climate politics were only about science and certainty, governments would use the IPCC process to speedily chart a way forward. But making the switch to low carbon and other greenhouse gas emissions will be a long and hard fight. It means changing the energy arrangements of the global economy, and provoking an industrial revolution as far reaching as its predecessor 200 years ago. It means using much less oil and coal, which are widely available and priced relatively cheaply, and using much more renewable power and clean fuels, which are currently in short supply and expensive. The economic and political stakes in making a successful transformation are very high, despite the massive potential for simultaneously cutting energy demand. In a world where facts are fought over like so many old bones by armies of editors, politicians, campaigners, columnists and corporate PRs alike, the truth takes a while to emerge and action that little bit longer still.

There are two main arenas where the arguments and choices are being resolved. The first and most obvious is the formal climate summits. They are showcase occasions, where what happens largely depends on progress made elsewhere. In these intense week-long events, governments negotiate hard, mostly in closed-door sessions, punctuated by the occasional open plenary for public posturing; and then they decide what will happen next. The media feverishly interpret the comings and goings for the watching world, report on progress and speculate about outcomes. Lobby groups from all sides attend as observers and advocates, briefing and opining in equal measure. The big set-piece days are the opening, when ministers arrive and positions are restated, and the close, when decisions are announced. The congress halls are decked out every day with hundreds of delegates, reporters and campaigners swirling around, trying equally hard to be heard and to hear the daily rumours and bargaining gossip. Outside, demonstrators provide colour, and poignant reminders of a wider concerned public looking on from a bemused distance. There is something of both the beanfeast and the bunfight to these annual jamborees. Recent summits – in Kyoto, where the protocol was first drawn up in 1997, in the Hague in 2000 where the talks collapsed, and in Bonn – have gone to the wire, with the outcome in the balance until late last-night deals are done. The process is sufficiently democratic and robust to ensure, so far, that violent protests have not featured.

In between summits, the debates and policy making continue at the nation state level. Given that it is governments who decide what happens at the summits, this second arena (or really arenas) actually matters more. The battalions of civil society ranged around the Convention's political territory are the politicos, journalists, campaigners and corporate lobbyists. Each side is well armed, with funds, staff, supporters and networks of allies. The opportunities to advance the cause for both sides are pretty much the same: policy making, publicity, argument and public opinion. The media news and analysis round-ups, Westminster and Whitehall, the high street and, occasionally, the courts, are the battlefields where they play. The prize is the policy position the nation state takes to a summit.

Each country's formal position reflects a number of factors. Strategic economic and foreign policy requirements, political ideology and national business interests are common to all. But, in the most influential and powerful nations, the rich, industrialized democracies, public opinion can still matter the most. The breakthrough at Bonn occurred because, in sufficient countries, a critical mass of public and professional opinion was more persuaded by the climate science and alternative energy economy prospects than by the do-nothing option of the antis. Voters and consumers were demanding action from their leaders and business. This was no accident, but the result of years of painstaking and focused campaigning by environmental organizations, most notably the three international groups (Friends of the Earth, Greenpeace and the World Wildlife Fund), across Europe, North America, Japan and Australasia. Their work mirrored and capitalized on the parallel scientific investigations of the IPCC.

No one was betting on a successful outcome at Bonn. The Climate Convention has a complicated formula for ratification, which includes securing the votes of at least 55 countries, that between them are responsible for at least 55 per cent of total emissions. The maths is designed to ensure that sufficient industrial countries sign up, as the main polluters, while guaranteeing sufficiently widespread participation from southern countries. Relatively fair, but easy to block. Worst still, Bonn was an emergency summit, called following the deadlock in the Hague, with an isolationist, pro-oil president newly occupying the White House.

At the Hague, the Clinton administration refused to sanction targets unless emissions could be offset against the carbon uptake of its forests. Without boring on about the many negotiating loopholes various blocs tried to secure for themselves, the 'forest sink' argument was the blockbuster. Key polluting countries like Australia and Canada wanted this concession; as did poorer nations with large forests, like Russia and Brazil, that would be able to sell pollution rights to other countries, notably America. The Southern nations, led by China and India, adamantly sought the economic and technical assistance and industrial nation cuts, due under the Convention. Japan's priority was to save international face and secure its first home-based (Kyoto) international treaty. To cut a long story short, Europe refused to concede a deal and the meeting ended in disarray. The logjam rolled over to Bonn. The battle lines were much the same. The major oil and coal producers like OPEC, the USA and Australia wanted business as usual to continue for as long as possible. A number of countries, led by Europe, wanted slowly to switch to cleaner sources of power, through convention-regulated target setting.

Public outrage, stoked by campaign groups, commentators and scientists, followed the collapse in the Hague, particularly in the industrial North. The European, Japanese and Canadian governments were under immense pressure to save the protocol, whatever the American position. On the second night after the summit was supposed to end, the Dutch chairman announced a deal had been done, to loud cheers. There were many concessions, including the lowering of Kyoto's already minimal reduction targets. Best of all, however, was the binding legal framework to cut emissions globally.

WINNING THE ARGUMENTS

Government is the flywheel of progress. Society does not deal well with too rapid economic, technological or social change. Governments slow things down. Political decision making is a paradox in that its horizons are invariably short term, yet, because there are so many layers of administrative and legislative procedure, and so many issues to address in a modern political economy, it takes a long time to change course fundamentally.

Low carbon energy strategies, based on renewable power, clean fuels and conservation, surfaced as a coherent alternative globally to the established fossil fuel and nuclear industry at the first Earth Summit in Rio. But with only tiny amounts of green energy then being generated, mostly by hydro, they were treated as theoretical and impractical options by decision makers. With eyes focused quite legitimately on the immediate task of ensuring enough power, heat and light, government and mainstream industry not unreasonably backed 'business as usual' energy policies. Campaigners and scientists who raised the thorny dilemma of rising greenhouse emissions back then were countered by officials who stated there was not enough evidence to act differently – and anyway where were the practical energy alternatives? Catch-22. Because without official support, green energy programmes were destined to remain on the drawing board.

The most difficult obstacle to overhauling an economic sector like energy or transport is developing the production capacity and distribution of replacement technologies to a critical mass, so that they can compete in the marketplace. It requires political leadership to change the governing rules to promote new supply and demand patterns. That has been missing up until very recently. Existing policies and entrenched market share give huge advantages to established companies using embedded methods that work to preserve the status quo. Fortunately, there are forces that push society to overcome the inertia of government and business. Since Rio, three social drivers have been instrumental in helping forge a new approach: new information and better science, developing markets for green energy technologies, and citizen action.

The IPCC investigations, and increasingly robust climate science, are leading to major energy and transport reviews in many countries. Legislative and executive arms of government are beginning to respond and slowly encourage solar, wind and wave power, clean fuels like hydrogen, and improved building and appliance efficiency standards. Accurate information convinces eventually.

Good information is another type of power, and democracies provide significant opportunities for aware citizens to influence public affairs. Environmental groups have been organizing both information streams and green consumers and voters, in order to pressure government and corporations to take climate change seriously and embark on safe energy paths. The Greenpeace campaign to stop new oilfields opening in the North Sea, which included the Brent Spar incident, and the Friends of the Earth (FOE) mobilization of hundreds of thousands of constituents across the European Union to ensure that member states backed the Kyoto Protocol in Bonn, are two examples from a host of similar actions undertaken since Earth Summit 1.

Information wars feature heavily on both sides. The two favoured tactics of the climate opposition, notably in the USA, Australia and Europe, has been to rubbish the science and warn of economic misery – soaring energy costs, bankrupt companies, lost jobs, falling market share and diminished household income – if legal reduction targets were ever adopted. In 1999, *The Ecologist* magazine painstakingly detailed the type of political influence and PR campaigns the sceptics have used to block green energy initiatives. As the long retreat from their science attacks accelerated, oil, mining and car companies paid lobby agencies, right-wing thinktanks and a handful of maverick academics to try to convince the public and politicians that only fossil fuels and free markets guarantee economic prosperity and liberty. Exxon/Mobil, The Heritage Foundation in the USA, the Institute of Economic Affairs in Britain and Burson-Marsteller are leaders. The *New York Times* and *The Guardian* newspapers have revealed the intimate political connections and lavish election funding between American oil and mining companies and the Bush administration.

Despite their investigative efforts, the media is a two-edged sword when it comes to improving public debate. Day to day, the complex issues which require nuanced analysis are submerged beneath the imperative to break news. Climate change-related events, such as increased petrol prices, floods, soaring traffic levels, IPCC reports or the summits themselves are just incidents, one potential story among many. Inevitably, the most dramatic aspects are highlighted to justify coverage, and a difficult issue becomes sensationalized simply by being abbreviated in editorial conferences, trailers and headlines. As today becomes yesterday, the race for tomorrow's news has already begun. From broadcast to broadcast, it is difficult to separate out the significant from the sound and fury. Over time, a different picture emerges, at least in societies where there is a free press and the internet. The repetitive nature of current affairs – where stories resurface, interviewers sift out reliable sources from the unreliable and new information links develop – overcomes the limitations of pulling together the daily million words or so of intelligible reporting that constitutes the average broadsheet.

In the end, it all comes down to economics and markets. When the economic countercase is strong enough, viable products appear in the stores and consumers feel confident, then political positions shift and policy reform follows. The market imperative is ultimately the most powerful lever for social change in consumerist societies. It can be stoked by analysis and campaigning, as environmentalists have realized. Groups such as the Worldwatch Institute and World Resources Institute in the USA, and FOE and Greenpeace internationally, have commissioned reams of authoritative research, economic modelling and case studies to demonstrate the substantial employment, competition and export opportunities of public and private sector investment in renewable supplies, public transport systems, clean fuels and energy conservation programmes. But to realize the benefits, policy makers have to create a level playing field in the market, and set targets for expanding the green economy and green technologies. Increasing the price of polluting fuels through high taxation and lowering the price of clean energy with low taxes is central to the green case. Withdrawing extensive public subsidies for fossil fuels and nuclear power, providing tax incentives for low carbon technologies, products and start-up companies, and incorporating environmental and social criteria in investment and trade rules are other necessary adjustments.

Consumer- and media-based campaigns run by groups like FOE and Greenpeace to persuade oil and car companies to go green have also had some noticeable effect, particularly in European-based energy giants Shell and BP Amoco, and the Ford Motor Company. Solar and renewable investments are up and the first hydrogen vehicles are anticipated by 2004. Some countries are committing to green energy futures. Iceland plans on becoming the first renewable energy economy within 30 years; with geothermal energy and hydrogen fuel cells powering vehicles, buildings and industry, including an extensive metals production sector built upon the world's cheapest electricity (see *World Watch* magazine, December 2000). Denmark is the world leader in building and installing wind turbines, because the industry has been nurtured by government policy. It generates substantial employment and export earnings. Business responds well to long-term strategic planning by government. Firms like to know what the score is. BP recently opted to build a new solar cell manufacturing plant in Spain, not Britain, because the Spanish government provided much better investment incentives. Decisions like this are convincing the UK that it needs to do more to support British green energy companies.

The Bush administration has determined that America's strategic interests are best served by backing oil and coal. US companies dominate these sectors globally, the domestic economy runs on cheap gas and the USA has huge fossil reserves. But this is the politics of the old economic order, and is untenable. In a global warming world, even a superpower like America will not be able to defend its fundamentally inequitable and unsustainable drain on the Earth's resources. The USA contains some four per cent of the planet's citizens, but annually these citizens consume 30 per cent of all oil production and release 25 per cent of carbon emissions. This ain't fair, and progressive Southern and Northern governments are demanding that the disparities end. The Convention process gives them the binding international framework to bring this about. Sooner or later, the isolationist republican strategy will deflate; as in the 1930s, when the party fought to avoid any involvement in European affairs while the Nazi spectre grew.

CONCLUSION

A wise sheik recently observed: 'The stone age did not end because they ran out of stones, and the oil age will not end because we run out of oil.' It is the conviction of new ideas that ushers in new eras, long before resources run out or old technologies cease to deliver.

The ecological necessity for low carbon economies is creating an enormous market for new power sources and energy management. The demand for heat, light, refrigeration and mobility is soaring all over the world, but is most needed in the poor South. In the North the sustainability challenge is to transform embedded infrastructures, plant and fuels; in the South, it is to create functioning energy economies.

Equity factors are influencing treaty decisions for the first time. Under the Convention, nations have agreed to distribute shares in the sky's capacity to safely absorb pollution, in the form of pollution quotas and reduction targets. Countries will trade in these shares, as the cap of overall pollution loads and individual quotas are progressively reduced. The financial value of energy investments will change relative to each other, as a result of these new market forces and public policy stipulations. Dirty power will become an expensive liability. Green energy will become more valuable to invest in, and cheaper to buy, for nations, companies and households.

For the transition to a low carbon global economy to occur, industrial quantities of renewable supply are required. Base-load electricity demand and clean energy to power hydrogen production are two needs. In many areas, including Britain, offshore wind, wave and tidal power are serious options. Fortuitously, many of the oil- and coal-rich nations, including America, Australia, Canada and the Middle East, are also blessed by the sun and wide open spaces and have the capital and technical resources to switch supply streams. Some large desert areas are almost certainly going to have to be used for mega-scale (many square kilometres) solar plants.

At the other end of the market, the climate conditions and lack of centralized energy infrastructures in many African, Latin American and Asian countries make their economies ideal development markets for mass-produced local and domestic solar products. As these multitudinous community needs are met, their living standards will rise as their emissions fall. As solar supply increases and production costs drop, these goods become more competitive in richer country markets. All the time, consumer demand for renewable energy and lower emission quotas will fuel similar green energy growth, jobs and services. Backed by appropriate incentives and penalties, such a transformation can be achieved within 25 years.

Whether this future happens or not depends, above all, on one political outcome: the strengthening and smooth functioning of the Convention. The biggest headache, after target setting, is how to allocate quotas. Currently, they reflect the basest political expediency – the rich, powerful, polluting countries get away with the lowest targets they can. Our representatives must learn a new politics, based on a very simple fact. There is only one sky. It cannot be made bigger. It is a global commons resource; an 'environmental space' of equal importance to every nation and everybody. It must therefore be shared on the same basis. Quotas and emission reduction targets should be set according to per capita needs, as well as ecological limits.

Doing so will give the South the quotas it needs to develop and a more equal political authority with Northern country power brokers. The lower quotas for the North provide a genuine incentive to innovate and improve economic performance through more efficient energy and resource use. Both North and South benefit. The convention process can deliver this outcome, by accounting for both ecological limits and projected population numbers as it distributes allocations on the way to an emissions cap of 60 per cent reductions by 2050. The just solution is the only workable one.

The idea of a global environmental space to be shared economically and politically by all people on an equal basis is a revolutionary concept. It gives rise to a world view as potentially compelling to our generation as those ideas of political economy encapsulated by Tom Paine, Adam Smith and Marx and Engels, which ushered in the industrial era, were to our forebears. It is pragmatic and visionary, and fits ecological realities, economic needs and political tensions in society. It anticipates a new international order, which limits economic competition within a framework of institutions and rules that encourage environmental protection, social equity and cooperation between all players – government, business and civil society.

As the new whole earth sciences teach us, humanity can no more escape the consequences of our dynamic interconnections with the natural environment than we can our entwined relationships with one another. This is an approach to living, working and playing on this planet that delivers for the individual and the community, and minimizes risk for generations to come. That is the prospect the climate debate has opened up for us. I believe it is our best future. It can happen, if enough voters and consumers demand it be so.

Essay Four

COPING WITH CHANGE: INSTITUTIONS FOR HUMAN HABITATION OF PLANET EARTH

Julian Morris

Climate change is nothing new. Climate has changed in the past, is changing now and will change in the future. But opinions about climate change have altered significantly. In the 1950s – a period of technological optimism – scientists advocated giant engineering projects in order to bring about change more rapidly. Today, by contrast, most scientific discussion of climate change is focused on the possibly negative effects of the actions of human beings, while public policy discourse is centred on the prevention of human-induced change. In this essay I will argue that this shift in approach, to emphasize the negative aspects of human activity and argue for preventing change, is highly problematic. Attempting to control the climate through restrictions on emissions of carbon dioxide and other greenhouse gases (GHGs) is not only undesirable, but is futile. As an alternative, I will argue that we should aim to structure society's institutions in a way that best enables man to cope with changes in the future. In particular, I argue in favour of creating decentralized political institutions, strong property rights and fair and effective legal systems. Furthermore, those political and legal institutions, charged with making decisions concerning the use of new technologies, should be constitutionally obliged to base their decisions on the best available scientific evidence, not on prejudice, fear or poorly substantiated claims of interest groups. Structuring our institutions in this way will ensure that resources are used more efficiently, that economies grow more rapidly and that beneficial new technologies are able to flourish.

◆ ● ●
● ●
● ● ◆ **IS TODAY'S CLIMATE OPTIMAL?**

Very little research has been conducted into the relative merits of different climates. Mostly, it is simply assumed that change is bad, even though there is evidence that some change may be beneficial. For example, global warming in the tenth and eleventh centuries made Europe more inhabitable, leading to population growth and improved living conditions.

Much is made by climate pessimists today of the possible adverse impacts of increased temperatures. It is said that hurricanes and floods will increase, devastating coastal regions and areas in floodplains; that sea-levels will rise, inundating small island states and low-lying regions such as Holland and Bangladesh; that tropical diseases will spread north and south as their vectors, such as the malaria-bearing anopheles mosquito, find the new climate more commodious.

But similarly apocalyptic predictions were made in the 1970s, when the conventional wisdom was that the world was likely to freeze over. Climate scientists such as Stephen Schneider and Lowell Ponte then claimed that excessive use of fossil fuels was leading to a build-up of aerosols in the upper atmosphere, blocking incoming solar radiation. The concern was that a small amount of human-induced cooling might trigger a vicious cycle leading to an ice age. In such a cooled world, deserts would expand as precipitation fell and crops would fail as growing seasons were foreshortened.

Global cooling might also exacerbate disease. There is anecdotal evidence that the great plague, which wiped out a third of the population of Europe in the fourteenth century, was exacerbated by global cooling. According to the historian Norman Cantor, crop failure (resulting from shorter growing seasons and less precipitation) led to malnutrition, which in turn made people more susceptible to disease.

From this we might conclude that all climate change is bad – and in a sense it is. All changes in climate pose problems for humans because we tend to establish agricultural practices and other facets of civilization (such as the location and structure of buildings) on the basis of particular climates. Likewise, climate change poses problems for other species that are genetically adapted for particular climates.

However, the mere fact that global climate change poses problems does not mean necessarily that we should attempt to prevent the climate from changing. It may, in the grand scheme of things, be more sensible to solve by other means the problems that arise. Consider some of the problems that might be associated with attempts to maintain a constant global climate.

PROBLEMS WITH ATTEMPTING TO ACHIEVE CLIMATE STASIS

First, since climate has changed considerably in the past of its own accord, we might expect it to continue to change in the future regardless of any increase in concentrations of greenhouse gases. So any policy aimed at controlling the climate must be capable of taking into consideration these natural climatic fluctuations. Failure to account for such natural variation might lead to a potentially catastrophic overreaction.

For example, some argue that most of the warming that has been observed in ground-based thermometers over the past 100 years is due to human emissions of GHGs. The same people typically claim that the world is already warming at an unsustainable rate. If that is true, then to achieve climate stasis purely by altering the concentration of GHGs in the atmosphere, would require a combination of radically reduced emissions (an immediate cut of approximately 60 per cent is said to be necessary) combined with an increase in carbon 'sinks' such as forests.

However, others argue that little of the observed surface warming is the result of increased atmospheric concentrations of GHGs. Rather, a combination of data bias (measured temperatures are biased upwards by various factors including urban expansion and land degradation), exogenous forcing (such as increases in solar radiation) and endogenous variation (the complex dynamics of the climate system itself) explain most of the change in observed temperatures. If these criticisms are correct, then the combination of reduced emissions and increased sinks could lead to a reduction in available carbon dioxide, which would have adverse effects for the growth of crops. Moreover, the restrictions on GHG emissions would have a devastating effect on human activity (see later).

Second, today's ecosystems reflect a dynamic process of adaptation to climate change that has persisted for millennia. If humankind succeeded in achieving climate stasis, the lack of change may actually be harmful to biodiversity, as those species that are particularly well adapted to today's climate would soon dominate their ecosystems, crowding out species adapted to different climates. Far from achieving the 'optimum' mix of species, the result would be a fragile and biologically barren world.

Third, there is likely to be considerable disagreement about what is the optimal climate for the world. Canadians, Scots, Norwegians and Siberians might well prefer a bit of GHG-induced global warming, as it would lengthen growing seasons and reduce the cost of heating in the winter. Meanwhile, people in Hong Kong, the Philippines, Florida and other tropical and sub-tropical areas might prefer a bit of global cooling if it would reduce the incidence and intensity of hurricanes/typhoons.

Our current understanding of the dynamics of climate change is inadequate to determine whether the incidence of violent storms will increase or decrease as a result of an increase in atmospheric GHG concentrations. In theory, the likelihood of occurrence of hurricanes increases as ocean temperatures increase. In practice, the evidence is at best inconclusive: there does not appear to be any correlation between hurricane numbers and temperature over the past 50 years. Some climatologists have even argued that because GHG-induced warming would occur primarily at high latitudes, the climate system would become more stable, resulting in fewer storms. In other words if climate stability *per se* is the objective, then GHG-induced global warming may be just what the doctor ordered!

ADAPTING TO CHANGE

Given the insurmountable problems with climate stasis as a public policy objective, it seems wise to seek an alternative solution to the problem of climate change. If, then, it is accepted that some change is inevitable, a key element of any climate change strategy must be adaptation.

For non-human species, adaptation means some combination of moving to places with more appropriate climates and genetically adapting to the changed climate. Human policy towards other species should focus on enabling those adaptive processes to take place as effectively as possible. That might mean, for example, creating 'wildlife corridors' through which species may traverse as climate changes. It might also mean genetically enhancing species to make them better able to cope with changed conditions. The latter strategy is particularly applicable to agricultural species, which have already been genetically modified for human use. To the extent that these domesticated species provide habitat to other, more 'wild' species, genetic enhancement will also have general benefits in terms of conserving biological diversity. In addition, higher yielding genetically enhanced domestic species may reduce pressure on the conversion of land for agriculture, thereby better enabling wild varieties to adapt. However, care must be taken to limit the extent to which genetic enhancements of domestic species cross over to wild species – as this also risks giving undue advantage to some species and thereby reducing species diversity.

For humans, adaptation will take many forms. In places susceptible to flooding, it might be desirable to construct higher levees or dikes. In places susceptible to typhoons or hurricanes it might be desirable to put in place early warning systems and to improve the construction of buildings. In some extremely low-lying areas, it might not be realistically possible to escape the rising seas and evacuation may be necessary. In other places, where the change in climate is less dramatic, the best strategy might be to insure against loss (and perhaps install air conditioning). It is clear that human adaptation will not take place of its own accord. All of these adaptive strategies require planning and financing. That leads to two questions. First, where will the resources come from to pay for the adaptive responses? Second, who will develop and implement the strategy for adaptation?

To the first question, one response would be: the 'polluters' should pay. But what if it turns out that most climate change is natural? Then there would be no polluter to pay, in which case those likely to be affected will have to pay – or convince others to donate to their cause. Regarding the second question, in most cases the person best able to make decisions regarding adaptation will be the individual, because it is the individual who must weigh up the benefits of investments in adaptation versus investments in, for example, next year's crop. In some cases, communities and even larger aggregations will be better placed to make decisions (for example where it is considered desirable to build levees or dikes).

ADAPTATION AT THE INDIVIDUAL LEVEL

At present, much of the world's population remains at the whim of nature. Hundreds of millions of peasant farmers experience periodic famines because they are not able to insure themselves against poor harvests. They live hand to mouth, with any surplus being sold on local markets at relatively low prices. Even those who are able to save for times of drought or flood may not be able to obtain adequate supplies because of inadequate distribution systems. The reasons for this tragedy are manifold, but at least one and usually a combination of the following key factors are usually present: lack of adequately defined property rights; inadequate enforcement of law; political control over domestic markets (of inputs and/or outputs); political corruption and instability; and restrictions on access to foreign markets. I now consider these factors in more detail.

PROPERTY RIGHTS

Clearly defined property rights are absolutely essential to economic development for two reasons. First, people with clear title to property know that they will be entitled to any returns that result from investments in improving their property, they have stronger incentives to make such improvements than people whose title to property is less clear. Second, clearly defined property rights enable owners to use their property as security against loans. Such loans tend to be available at lower rates than unsecured loans, thereby reducing the cost of investing in, for example, yield-enhancing new crop varieties, fertilizer and pesticide. To peasant farmers, such investments offer the opportunity to escape from subsistence, save and thereby to insure against poor harvests. Alternatively, such loans may be used to develop a business other than farming. Such entrepreneurial activity is an essential adjunct to improving farm output, as it offers people another means of insuring against an uncertain world. In addition, it increases labour productivity, as seasonal farm labourers are able to work in other jobs during the off-seasons.

THE RULE OF LAW

Clearly defined property rights are of little use if they cannot be enforced. Such enforcement should ideally occur through the rule of law – the non-discriminatory application of rules relating to property, contract and other legal relations. When everyone is equal before the law in this way, people are more willing to engage in entrepreneurial activity.

FREE MARKETS

However, in order for businesses to flourish, not only is it necessary for property rights to be defined and enforced and for contracts of sale to be upheld, but people must also be free to enter into contracts unhindered (as far as possible) by state intervention in the form of red tape and taxes.

POLITICAL STABILITY

War and terrorism do not make the ideal environment for a burgeoning economy. They tend to interfere with the protection of property and the enforcement of contracts. Likewise, the risk of war and terrorism undermines the incentive to invest in the development of all forms of business enterprise. Political stability is thus a prerequisite for economic development and adaptation to change.

ADAPTATION AT HIGHER LEVELS OF AGGREGATION

While improved economic circumstances at the individual level are essential for the poorest in the world to be able to adapt to change, economic growth also enables more to be spent on adaptation at other levels of aggregation. Villages threatened by floods will be able to spend more on building levees to defend themselves. Individuals, companies and communities will also be able to spend more on defence against disease. Demand for clean water will rise and so the incidence of water-borne infections, such as hepatitis, is likely to fall. Similarly, spending on programmes to limit infestations by disease-carrying insects, such as the tsetse fly and the anopheles mosquito, is likely to increase with commensurate reductions in the incidence of associated deadly diseases. Meanwhile, spending on vaccination against and treatment of communicable diseases such as TB is also likely to rise, leading to a reduction in the rate of such diseases. Finally, spending on education programmes is also likely to increase, leading to more responsible behaviour with regard to many diseases, including AIDS.

Private individuals or charities may enact some of these measures and local, regional and national governments may enact others. What kind of organization takes responsibility is not automatically obvious, neither is the level of investment. Where government does get involved, there will be a tension between investments in adaptation in the short term and enabling investments in adaptation at all levels in the long term; inasmuch as spending on adaptation at the regional and national level requires taxation, such spending will reduce individual income levels and slow down economic growth. This tension can best be resolved by devolving responsibility for making decisions concerning adaptation to the lowest level possible and ensuring that whoever is making these decisions is accountable for their actions, either through contract or through some system of democracy.

DECISIONS REGARDING NEW TECHNOLOGIES BASED ON SOUND SCIENCE

New technologies are a key driver of both economic development and adaptation to changing environments. Some such new technologies enable increased resource efficiency and thereby increased economic output. Others, such as GM crops, may enable more effective adaptation (as well as increasing efficiency by reducing the need for other inputs). Already GM crops have been developed that are able to withstand hotter, drier, and saltier environments. However, opposition to new technologies in rich countries threatens to undermine their use globally. Already, governments of many poor countries have imposed restrictions on the use of GM technologies because of fears that GM products will not sell well in Europe.

While some new technologies may have adverse consequences – and therefore should be regulated – the majority of new technologies provide benefits to mankind far in excess of their costs. To put it another way, new technologies may entail new risks, but they tend to reduce the risks we face both from older technologies and costs more generally. Gas-fired central heating entails risks of carbon monoxide poisoning and explosions, but is generally less dangerous than coal-fired stoves (which typically emit a cocktail of noxious gases and particulates). But coal-fired stoves were an improvement over poorly flued coal or wood fires, which were even more noxious and entailed an even higher risk of death from fire. And wood fires were an improvement over no fire at all; fire provided warmth, preventing hypothermia and also enabling our ancestors to cook food, reducing the incidence of food-borne bacterial infection.

The imposition of swingeing restrictions on the use of new technologies may prevent some unforeseen consequences of those technologies, but it will also reduce our capacity to deal with other unforeseen harms. Regulatory decisions should be based on the best available scientific evidence and should balance the likely benefits from the new technologies against the often overemphasized and typically hypothetical costs.

FREE MOVEMENT OF PEOPLE

In addition to these key criteria a number of other, less important, criteria are also important. For some people living in places that might be subject to severe climate change, the most appropriate solution to change might be to move. It is thus essential that those people be permitted freedom of movement.

ADAPTATION IN CONTEXT

In combination, these institutions will ensure that people are better able to cope with climate change, if that is their concern. For poor people, however, more immediate concerns may dominate. At present, hepatitis, TB, AIDS and other communicable diseases present a far more deadly threat to people in Africa, Asia and South America than does climate change. Likewise, water-borne diseases such as amoebic dysentery, and vector-borne diseases such as malaria, leishmoniasis and schistosomiasis continue to kill millions of children. A combination of economic development and targeted healthcare programmes offer the best hope of reducing such tragedies. Until death rates from such problems are dramatically reduced, the idea of spending resources on limiting climate change seems morally obscene.

ALTERNATIVES: LIMITING CLIMATE CHANGE

While adaptation is a necessary component of any strategy to cope with climate change, there may also be a place for policies to alter the nature and extent of change more directly. Limiting GHG emissions may be one way to achieve this. But before we rush headlong into a programme of GHG reductions, it is worth considering the implications of such a programme and some of the alternatives.

The current proposed scheme, agreed at Kyoto in 1997, requires rich countries to reduce their emissions of six GHGs by an average of five per cent below 1990 levels by 2008–12. If implemented, this would have only a small effect on global concentrations of carbon dioxide and a smaller effect on climate. If the predictions of the Intergovernmental Panel on Climate Change are correct, the world would be perhaps 0.15°C cooler in 2100 than if no restrictions had been imposed – and that is not taking into account offsetting increases in emissions from poor countries, which have not agreed to any reduction. Such a change would probably not be distinguishable from natural variation. Meanwhile, the cost to rich countries would be significant. Even the most optimistic predictions estimate that the cost would be about 0.1 per cent of GDP in OECD countries. Taking into account the inefficient way in which many countries are planning to implement Kyoto (for example by subsidizing 'alternative' technologies) and the lack of acceptance of trading of carbon emissions permits by many governments, the annual costs are likely to be closer to 0.5 per cent of GDP or US$125 billion. That cost will feed through into slower growth globally, with adverse effects especially on people in poor countries. And for no real benefit!

To see how ridiculous the Kyoto Protocol is, just think what else that money could have been spent on. It is estimated that a mere $6 billion annually, if spent wisely, could save hundreds of millions of lives from infectious disease. That is less than five per cent of the loss in output. Assuming an average tax take of 30 per cent in OECD countries, that means that if one fifth of the tax that is lost as a result of implementing Kyoto were otherwise spent on providing medicines to people in poor countries, millions of lives would be saved each year. Likewise, if only five per cent of that $125 billion was spent on goods and services from poor countries, people in those countries would be more than $6 billion better off – both in pure material terms and consequently in terms of their ability to adapt to a changing climate.

OBJECTIONS

Some people counter by saying that we use energy inefficiently and that by imposing restrictions on the use of carbon-based fuels and by subsidising 'alternative' energy, society will benefit. While it is true that there are inefficiencies and that the market is not perfect, most of the inefficiencies are the result of government intervention (such as subsidies to coal production, restrictions on the use of natural gas and taxes on oil). Meanwhile, attempting to correct for market imperfections typically results in cures that are worse than the problem – while leaving the problem itself more or less unchanged.

Consider, for example, energy efficiency in homes. It is often contended that homeowners under-invest in energy efficiency expenditure. Responding to these concerns (and to demands from the producers of thermal insulation), the British Government has mandated that new houses must meet certain thermal conduction criteria. However, much of the housing stock in Britain dates from a period when insulation was not considered particularly important. The rules on new home insulation have no effect on this part of the housing stock. Meanwhile, planning and building regulations prevent the replacement of the old stock with more energy-efficient new stock. In some cases, planning regulations even prevent the replacement of draughty old windows with heat-containing double-glazing. If people were given the freedom to decide what to build and where (subject to reasonable constraints), they would no doubt replace much of the ugly, dark, damp stock of British housing with something rather more pleasant – and more energy efficient.

It is clearly desirable to remove government-created inefficiencies in the energy market (such as taxes, subsidies and unnecessary regulations). The removal of such distortions would have both environmental and economic benefits. But the imposition of restrictions on emissions of GHGs, either directly or by the proxy of taxes on carbon-based fuels, would drive up the cost of energy, which would make entrepreneurial activity more difficult and generally harm the economy, leading to unemployment and misery.

The common assertion, that subsidies to alternative energies would create jobs, is economic nonsense. Subsidies to alternative energies mean either an increase in energy prices (if, as in the UK, alternative energies are subsidized by a levy on carbon-based energy) or an increase in taxes. Either way, overall output will fall as costs of production rise. That means people will lose their jobs. Yes, a few people may be employed building and managing wind turbines, but many people will lose their jobs in everything from steel mills to internet companies. (As the recent experience with energy shortages in California demonstrates, even the high-tech world of the internet requires electricity.)

A second objection is predicated on the assumption that the imposition of a tax on GHG-emitting forms of energy generation can be offset by a reduction in other taxes. While there is some merit in the idea that all factors of production (energy, labour, capital) should be taxed at the same rate, the fact is that economic activity is already massively overtaxed in nearly every country in the world, so any new tax should be treated with scepticism. Public opposition to new taxes can be vehement – as with the poll tax riots in London in the late 1980s – and governments are therefore increasingly looking for less visible ways of taxing the public. If government can justify a tax on the basis that it will help the environment, it is really on to a winner. But if the tax will not help the environment and will simply be used to provide more inefficiently supplied 'public' services, the public are right to oppose it. In an era in which government seems to have accepted the proposition that it is incapable of efficiently supplying almost any service (in Britain failing state schools are being run by private companies; waste services are the subject of compulsory competitive tendering; even the railways were eventually privatized – albeit in a cack-handed way), it is surely time to start reducing both expenditure and taxes.

Another common contention is that by reducing the use of carbon-based fuels we are helping future generations. This argument is based on the flawed assumption that such fuels are being used at an 'unsustainable' rate. This assumption in turn is based on the anecdotal observation that at any point in time we seem to have only a fixed number of years' worth of supply of any particular resource. In fact these fuels remain available in abundant quantities in spite of high rates of extraction for over a century. The reason that at any point in time there is only 30–50 years' worth of supply of oil and gas has to do with the commercial repayment period for investment in discovery of new sources. If, at some point in the future, the costs of extraction increase sufficiently, then energy production companies will begin to diversify their portfolio by investing in alternative energies. Most current investments in solar and wind power are a means of acquiring government subsidies. Most of the rest can be explained as public relations exercises. A very few may make sense – as sources of electricity in areas too remote to have distributed electricity; but even these may be uneconomic compared to technologies such as pebble bed nuclear reactors.

RADICAL CARBON CONTROL

Kyoto, one might conclude, is deadly. But more swingeing restrictions would be even deadlier. Amazingly, however, many environmental groups actually demand restrictions of up to 60 per cent.

Even if it were politically feasible to institute such a programme of restrictions on the use of carbon-based fuels, the human cost would be enormous. For people in the rich world, the cost would be felt in terms of rising unemployment and declining levels of self-worth. In some places, especially in areas where there had been significant concentrations of heavy industry, there may be social unrest. Many people would no doubt commit suicide as they saw any hope for the future disappear.

For people in poor countries, the impact would be devastating. Many people just now rising from the servility of subsistence farming would essentially be forced back to the land to scrape a living – or, in many cases, to die. With food more expensive and less plentiful – because of the increased cost of factor inputs (especially seed, fertilizer, pesticide and petrol for tractors) – malnutrition would be widespread. With weakened bodily defences, people will more easily succumb to disease. This will be exacerbated by cuts in government expenditure, which will result in reduced spending on programmes to control diseases such as TB, hepatitis, malaria and AIDS.

Realizing the political difficulties of suggesting that poor countries should cut back emissions, some groups have suggested that rich countries should cut back emissions by up to 90 per cent – in order that poor countries may increase their emissions. The ideological motive behind this suggestion is that all persons should have the right to emit the same amount of GHGs. Not wishing to get too deep into the dubious philosophical underpinnings of such a proposition, it is sufficient to observe that if the rich countries did reduce their emissions by 90 per cent, then barring a technological miracle, the impact on poor countries would be so devastating as to render irrelevant their nominally permitted level of emissions. Trade with rich countries is an important part of the process by which poor countries will hopefully ultimately cease to be poor. Making rich countries poor would solve the rich–poor divide by mutual suicide.

◆ ● ●
● ●
● ● ◆ **CONCLUSION**

Ten years ago, in his presidential address to the American Economics Association, Thomas Schelling noted that the policy maker of 1891 looking forward a century would perhaps have seen horse manure as the most pressing environmental problem. Indeed, if as many people used horses to drive into London as today drive cars and other forms of motorized transport, London would be many feet deep in excrement. But technologies change and the problems associated with them change too. Today's obsession with controlling carbon dioxide is likely to look, 100 years from now, very much like the problem of horse manure might have appeared to a policy maker a century ago.

The people of the world face a long list of genuine problems. Climate change is very low down on that list. We should not pretend that it is more pressing and we should be very wary of imposing a 'solution' to the problem that is worse than the problem itself.

Any proposal to address the problem of climate change should be considered in light of the consequences – intended and unintended – of enacting such a proposal. I have argued that the current proposals to limit GHG emissions would have costs that exceed their benefits. In particular, rates of economic development in every part of the world would be reduced. As a result, the poorest people in the world would be less able to adapt to the changes that will inevitably occur.

Humans are by their nature an adaptive species. We have adapted to past changes and we will adapt to future ones. It would be tragic if by our attempts to control the uncontrollable climate we were to undermine our ability to adapt to unforeseeable future changes.

AFTERWORD
Tony Gilland

All contributors to this book agree that global warming is taking place and that, particularly in the last 30 years, the effect of rising concentrations of human-produced greenhouse gases have most likely contributed to this warming. However, there is disagreement among the authors first as to the significance we should attach to climate change; second regarding policy responses to global warming; and third on the wider issue of humanity's relationship to the natural world.

HOW SIGNIFICANT IS CLIMATE CHANGE?

Mike Hulme warns that climate change is of great significance. A change of even 1° Celsius, he contends, will alter 'the distribution and magnitude of precipitation over the earth's surface' and some potentially serious disruptions may result, such as greater risk of flooding 'or changes in the distribution of certain disease vectors and pathogens.' Charles Secrett goes further, stating that the threat of serious climate change has imposed 'a planet-wide array of interlocking problems on society; and a set of choices of unparalleled magnitude.' In contrast, Peter Sammonds argues that the key question is whether there is going to be a benign warming of up to 2 to 3°C or whether the warming will be more extreme. According to Sammonds, the high end of the IPCC's prediction of temperature increases of between 1.4 and 5.8°C by the end of 2100 is 'a consequence of the double application of the precautionary principle both in regards of global socio-economic development and how the climate system works.' That is, the outcome of one worst case scenario has been multiplied by that of another, to arrive at the higher predictions for global warming. Julian Morris argues that because the climate can change considerably on its own accord, attempts to control for human impacts upon the climate run the risk of creating 'a potentially catastrophic overreaction' if these natural variations are not properly taken into account. Given the scientific uncertainties involved he suggests this is a real possibility.

It is worth noting that the opinions of Hulme and Secrett are closest to those that hold sway over the public debate in Europe and to the perspectives adopted by most European governments. This situation has led to high-profile international discord between the European Union and the Government of the USA.

RESPONSES TO CLIMATE CHANGE

Policy responses to climate change advocated in this book fall into two main areas.

BETTER SAFE THAN SORRY?

If it is accepted that emissions of greenhouse gases resulting from human activities are contributing, at least to some extent, to climate change one response is to argue that we should take steps to reduce human impacts on the global climate. The possibility of any harmful consequences of global warming occurring can be minimized this way, it is argued. Governments that are acting upon and supporting the implementation of the Kyoto Protocol have taken a step down this route by taking actions to reduce the emission levels of their countries. Such an approach is in line with the emphasis on the need for caution in the face of uncertainty – a principle that has become increasingly influential in policy circles over the past ten years.

However, within this broad policy response there are differences of opinion as to how much caution should be exercised – particularly between different national governments negotiating the terms of the Kyoto Protocol and between those governments and environmentalists. Charles Secrett is clearly pleased with the progress made over 2001 towards ratifying the protocol. However, he also views it very much as a first step. As he points out, if the targets that have been set are achieved, this will only reduce carbon emissions by a few percentage points below 1990 levels by 2012. This will leave much to be done if the 60 per cent reductions that the IPCC argues are required by 2050 to stabilize the impact of emissions on the climate are also to be met. Secrett argues that renewable energy supplies are the way forward and will not necessarily require less economic growth since they would provide substantial employment, competition and export opportunities.

PROMOTE ECONOMIC GROWTH?

Taking the example of the terrible floods in Mozambique in 2000, where peasant farmers lives were devastated, Sammonds argues, in stark contrast to Secrett, for the 'rapid industrialization and the application of advanced technologies for Mozambique.' For Sammonds the key issue in this instance was insufficient financial resources to provide comprehensive flood defences and to cope with an agricultural crisis. The lesson he draws is that we must not 'check development through a misplaced sense of what is appropriate for the country' and that more generally 'nothing should be done which would curtail global economic growth through misplaced strategies to curb carbon dioxide emissions.' This approach differs from the first in that it suggests that attempts to minimize our impact on climate change will divert our resources from the more important task of promoting strong economic growth which will enhance our capabilities to cope with any future changes in the climate that may occur.

Morris is also concerned with economic development. He rejects attempts to control carbon emissions as futile and argues instead for clearly defined property rights, the rule of law, free markets and political stability to provide the conditions most conducive to economic growth. Morris is dismissive of the idea that alternative (renewable) energy is currently an economically viable way forward. He argues that the present requirement for subsidies to alternative energies 'mean either an increase in energy prices or an increase in taxes. Either way, overall output will fall as costs of production rise. That means people will lose their jobs.'

BROADER ISSUES

The debate about how we should respond to the prospect of climate change is not simply a practical one about policy responses. It is also concerned with the broader issue of our relationship with and attitudes to the natural world. Differences of opinion on how we should relate to nature lead to contrasting views about what constitutes responsible and irresponsible behaviour on the part of individuals, institutions and governments.

NEW MORAL PROBLEMS?

Some argue, for example, that our awareness that human activity can affect something as important and fundamental as the global weather system means we must draw important conclusions about our wider relationship with nature. In particular, it is often suggested that humanity should be more humble in the face of the complexities of the natural world and the unforeseen impacts human behaviour can have on the planet. From this perspective we need to reassess what have been considered to be great human achievements, such as industrialization, and question if they were so beneficial after all. From governments down, we need to fundamentally modify our behaviour. Secrett highlights, in this vein, what he sees to be important new moral dilemmas for the individual, and argues that 'everyone, not just government and industry, has to play a part.' According to Secrett, 'our lifestyles and the pollution we cause, spark economic and social consequences which stretch across the global community and for future generations' and this opens up 'an unfamiliar moral dimension to everyday decisions, ones usually taken without a second's thought.' This perspective has become increasingly influential in modern society and is reflected in the British Government's current poster and television advertising campaign which asks people: 'Are you doing your bit?' [for the environment].

MORALISM IS THE PROBLEM

For some the situation can be understood quite differently. We should recognize and view positively the ingenuity and capabilities of humans to generate and respond to change and be prepared to deal with unforeseen consequences in the future having expanded the opportunities and options before us. From this perspective, the view that connects global warming to the issue of responsible and irresponsible behaviour is unhelpful. For Sammonds, the debate

NATURE'S REVENGE? Afterword

about global warming has adopted a 'strongly moralistic tone' whereby upholding the view that global warming 'poses a terrible threat to the planet ... is clearly now a measure by which the morality of nations will be judged.' He links this moralizing to the loss of a 'bold humanistic mission' that was 'outward looking and viewed risk taking with some optimism.' According to Sammonds, such moralizing has a negative impact on the quality and seriousness of the public debate and can even distort the discussion of the science. For example, he argues that it has led many pundits and politicians to take for granted that a warmer climate will lead to more storms and droughts despite the fact that the IPCC found that observed variation of these events shows 'no significant trends over the last century.' From this perspective, climate change has been wrongly endowed with the significance of a moral tale about human folly, which clouds serious debate and denigrates the positive role economic and social developments have played in improving our lives.

At the heart of this debate are important and fascinating scientific questions about the processes of climate change. However, as the essays in this book illustrate, political and moral questions about how we view ourselves and the future development of society play an equally, if not more, important role in the debate. We hope this book has stimulated your thoughts on this important issue.

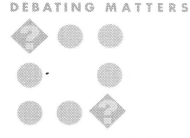

DEBATING MATTERS

Institute of Ideas
Expanding the Boundaries of Public Debate

If you have found this book interesting,
and agree that 'debating matters', you can
find out more about the Institute of Ideas
and our programme of live conferences and
debates by visiting our website
www.instituteofideas.com.
Alternatively you can email
info@instituteofideas.com
or call 020 7269 9220 to receive a full
programme of events and information about
joining the Institute of Ideas.

Other titles available in this series:

DEBATING MATTERS

Institute of Ideas
Expanding the Boundaries of Public Debate

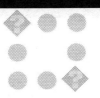

SCIENCE:

CAN WE TRUST THE EXPERTS?

Controversies surrounding a plethora of issues, from the MMR vaccine to mobile phones, from BSE to genetically-modified foods, have led many to ask how the public's faith in government advice can be restored. At the heart of the matter is the role of the expert and the question of whose opinion to trust.

In this book, prominent participants in the debate tell us their views:

- Bill Durodié, who researches risk and precaution at New College, Oxford University
- Dr Ian Gibson MP, Chairman of the Parliamentary Office of Science and Technology
- Dr Sue Mayer, Executive Director of Genewatch UK
- Dr Doug Parr, Chief Scientist for Greenpeace UK.

COMPENSATION CRAZY:

DO WE BLAME AND CLAIM TOO MUCH?

Big compensation pay-outs make the headlines. New style 'claims centres' advertise for accident victims promising 'where there's blame, there's a claim'. Many commentators fear Britain is experiencing a US-style compensation craze. But what's wrong with holding employers and businesses to account? Or are we now too ready to reach for our lawyers and to find someone to blame when things go wrong?

These questions and more are discussed by:

- Ian Walker, personal injury litigator
- Tracey Brown, risk analyst
- John Peysner, Professor of civil litigation
- Daniel Lloyd, lawyer.